SpringerBriefs in Computer Science

T0253975

For further volumes:
http://www.springer.com/series/10028

David Gerónimo · Antonio M. López

Vision-based Pedestrian Protection Systems for Intelligent Vehicles

 Springer

David Gerónimo
Antonio M. López
Computer Vision Center
Universitat Autònoma de Barcelona
Bellaterra, Barcelona
Spain

ISSN 2191-5768   ISSN 2191-5776   (electronic)
ISBN 978-1-4614-7986-4   ISBN 978-1-4614-7987-1   (eBook)
DOI 10.1007/978-1-4614-7987-1
Springer New York Heidelberg Dordrecht London

Library of Congress Control Number: 2013940300

Printed on acid-free paper

Springer is part of Springer Science+Business Media (www.springer.com)

Preface

This book is a natural evolution of the survey paper we published in the IEEE Transactions on Pattern Analysis and Machine Intelligence [121]. Therefore, we have followed the same point of view. On the one hand this means that we focus on vision-based solutions for pedestrian detection. Moreover, although we include some references from fields such as surveillance, robotics, and multimedia, the backbone context of the book is driver assistance, i.e., the development of pedestrian protection systems. On the other hand, we have organized the literature according to the main stages that a vision-based pedestrian protection system must incorporate. Moreover, we do not focus on the mathematical formalism behind each proposal. Rather, we try to explain the overall ideas so that interested readers can go to the original references to look into the details. Because of that, we think that novel researchers in this field, from the academia or the industry, can take benefit of this book. At this moment, [121] has received already more than 140 cites. We have informally collected feedback from some readers of [121] and we are glad to say that it seems that this paper accomplished its purpose, namely to give a quick overall organized idea of the field to novel researchers and even to experienced researchers of other fields (e.g., it was interesting to discuss with researchers of the machine learning community about detection vs classification). Thus, we hope that in the same line this book can be useful even for more researchers.

In this book more than 300 papers are referenced, which gives an idea of the enormous interest that vision-based pedestrian detection has deserved from both the computer vision and the intelligent transportation systems communities. We include more than 190 references that were not included in [121], more than 80 published after [121], which shows how active this field is. In spite of the amount of papers we reference, there are also many more that have not been included. We apologize for that, since all published works deserve their own acknowledgment given the effort of the authors. However, even a book like this one needs to limit the space in order to keep on focus. We do not provide a sort of best papers ranking, but we have chosen those papers that better allow us to illustrate the diversity of solutions available in the literature. Of course, this implies that papers that presented breaking ideas along this past decade are included.

It is worth to mention that vision-based pedestrian detection is a subfield of object detection in videos, which in turn deserves its own book given the enormous corpus of proposals generated so far thanks to advances in topics such as machine learning, image descriptors, computational power, etc. This means that some general object detection methods can be applied to pedestrian detection. However, we have focused on the literature that explicitly addresses pedestrian detection in its own. As we will see, modern pedestrian/object detectors mainly rely on image descriptors (features), pedestrian/object models (e.g., holistic, multi-aspect, part-based, etc.), and learning machines (e.g., SVM and AdaBoost variants). In this book we focus our explanations on the content more specifically related to vision, i.e., pedestrian image descriptors and models. While the machine learning algorithms used in the reviewed literature are of course mentioned, explaining how such algorithms work is out of the scope of the book.

This book is not done only thanks to the work of the authors, but many more persons and institutions must be acknowledged. First, we thank to the pedestrian detection community itself for addressing such a challenging and socially relevant problem. We thank also Springer for giving us the opportunity of writing this book. We thank to our daily collaborators at the Computer Vision Center (CVC) of Barcelona as well as at the Department de Ciències de la Computació of the Universitat Autònoma de Barcelona (UAB). Special thanks to the administrative and support staff who makes research easier. Many thanks to all the members of the CVC-ADAS group (www.cvc.uab.es/adas) for the hard work along the 10 years of life of the group and for being people that we really enjoy to work with. Special thanks also to David Váquez, Javier Marín, and Jiaolong Xu for kindly helping us to explain better some parts of this book. We want to give thanks also for the public funding we have received along the last years to support our research in the driver assistance context. In particular, we thank the following projects from the Spanish Government: TRA2011-29454-C03-01, TIN2011-29494-C03-02, TIN2011-25606, TRA2010-21371-C03-01, Consolider Ingenio 2010: MIPRCV (CSD2007-00018), TRA2007-62526/AUT, and TRA2004-06702/AUT.

Antonio M. López also wants to thank his Ph.D. supervisor Joan Serrat, as well as mentors Juan José Villanueva, and Bart M. ter Haar Romeny. Good supervisors and mentors are key for having a successful career, these boys are. Equally important is to have good Ph.D. students who help you to stay updated and active in the frontier of the research. Accordingly, thanks also to Antonio's Ph.D. students Daniel Ponsa, David Gerónimo, José M. Álvarez, David Vázquez, Javier Marín, Diego Cheda, Yainuvis Socarrás, Muhammad Rao, Jiaolong Xu, and Sebastián Ramos. In addition, many thanks also to researchers who have hosted these students during researcher stages and with whom I have enjoyed research and life discussions. Thus, many thanks to Theo Gevers, Krystian Mikolajczyk, Ludmila Kuncheva, Dariu Gavrila, Bastian Leibe, and Frédéric Lerasle. Thanks also to other CVC-ADAS Ph.D. students with whom I have actively collaborated as Carme Julià, Ferrán Diego, José C. Rubio, and Germán Ros. As head of the CVC-ADAS team, Antonio also wants to give special thanks to the members who fought to build the team from the scratch, namely Joan Serrat, Felipe Lumbreras, Angel

D. Sappa, and Daniel Ponsa. Antonio wants to thank also the private companies that trusted the CVC-ADAS team to develop some ADAS-related projects, in particular, thanks to Volkswagen A.G. at Wolfsburg (Thorsten Graf, Jörg Hilgenstock) and SEAT Centro Técnico of Martorell. Finally, Antonio wants to apologize with the family, specially with his parents, Antonio and Iluminada, his brother Juan, and specially with his wife Ana for stolen so many time supposed to be free-time for attending the research, many thanks for the enormous patience and support.

David Gerónimo wants to thank his former Ph.D. supervisor and currently research collaborator Antonio M. López for the guidance and support during the last 8 years. Day by day, his attitude toward research has been an inspiring source for him to improve as a researcher. Many thanks also to his collaborators, research fellows and friends for the long and fruitful discussions: Angel D. Sappa, Joan Serrat, Daniel Ponsa, José M. Álvarez, David Vázquez, Javier Marín, Didier Devaurs, Hirofumi Uemura, Alejandro Mosteo, David Aldavert and Xuan Zou. Thanks also to David's hosts during research stays in Surrey and LAAS-CNRS, Krystian Mikolajczyk and Frédéric Lerasle, for their warm welcome to their labs. Finally, David wants to thank his family, specially to his parents Angel and Núria, and his sister Sara, for their support. Finally, David's most sincere acknowledgments are addressed to his wife Verónica, for her understanding and never-ending support, specially during the long and untimely hours spent in research.

Contents

1 Introduction 1
  1.1 Automobile's Impact 1
  1.2 Advanced Driver Assistance Systems.................... 4
  1.3 Pedestrian Protection Systems......................... 5
  1.4 The Role of Computer Vision.......................... 7
  1.5 Generic Framework 8
  1.6 Book Outline 9

2 Candidates Generation................................... 13
  2.1 2D-Based Approaches 13
  2.2 3D-Based Approaches 17
  2.3 Motion-Based Approaches 20
  2.4 Discussion 20

3 Classification 23
  3.1 Preliminary Concepts............................... 24
    3.1.1 Image Descriptors 24
    3.1.2 Pedestrian Models 25
    3.1.3 Pedestrian Classifiers........................ 26
  3.2 Holistic Models: Focus on the Features 26
    3.2.1 Templates................................. 26
    3.2.2 Haar Features 30
    3.2.3 Edge Orientation Histograms Features 32
    3.2.4 Histogram of Oriented Gradients Features 33
    3.2.5 Shapelet Features........................... 36
    3.2.6 Local Binary Pattern Features 36
    3.2.7 Dominant Orientation Template Features 38
    3.2.8 Co-Occurrence Features 39
    3.2.9 Covariance Features......................... 42
    3.2.10 Data-Driven Features 43

3.3   Diversified Models: From Features Fusion to Multiple Parts. . . .   45
     3.3.1   Combined Features   45
     3.3.2   Classifier Cascades/Ensembles   48
     3.3.3   Multiple Aspects   49
     3.3.4   Multiple (Body) Parts.   52
     3.3.5   Multiple Resolutions   57
     3.3.6   Occlusion Handling   60
3.4   Training   62
     3.4.1   Parameters Tuning.   62
     3.4.2   Bootstrapping   63
     3.4.3   Data Annotation   65
     3.4.4   Domain Adaptation   67
3.5   Discussion   68

4   **Completing the System**   73
4.1   Preprocessing   73
4.2   Verification and Refinement   76
4.3   Tracking   78
4.4   Application   80
4.5   Real-Time   81
4.6   Discussion   83

5   **Datasets and Benchmarking**   87
5.1   Datasets   87
5.2   Evaluation Protocols   90
5.3   Discussion   92

6   **Conclusions**   95
6.1   State of the Research   95
6.2   Future Challenges   97

References   99

Chapter 1
Introduction

Intelligent machines assist us in many situations of our everyday lives: medical diagnoses, communications, transportation, education, surveillance, etc. It is expected that during the current century machines will become even more integrated in our daily experiences, especially focused on improving aspects such as safety and comfort.

This book describes a type of systems aimed at avoiding pedestrian to vehicle collisions. These systems, formally known as Pedestrian Protection Systems (PPSs), must detect and track pedestrians, and provide the necessary outputs to the host vehicle in order to prevent potential accidents and even reduce their severity when unavoidable. The breakthrough of these systems occurred at the beginning of the twenty-first century thanks to the advances in sensing, the maturity of artificial intelligence and computer vision, and the increase in machines' computational power. In the last years intensive research efforts in this technology have been carried out by both public and private entities, which have projected their massive commercialization during this decade.

1.1 Automobile's Impact

When people are asked the most relevant technology that has changed most the landscapes of our cities during the last century, the automobile is a common answer. Indeed, the human development in the modern era is represented up to some extent by the automobile. It has changed societies not only in urban planning, industry and economy but also in demographic distribution and social interactions. Employment, leisure and relationships are all shaped to a greater or lesser degree by automobiles. This is clearly illustrated when comparing two families from the early 1900s and the 2000s. The former family bought their food and clothes in local stores, close to the living place, traveled to nearby working places, and the leisure travels were restricted to the nearby regions. Nowadays, most of the shopping activities are concentrated in

D. Gerónimo and A. M. López, *Vision-based Pedestrian Protection Systems*
for Intelligent Vehicles, SpringerBriefs in Computer Science,
DOI: 10.1007/978-1-4614-7987-1_1, © The Author(s) 2014

large shopping malls out of the cities, working places are farther away and the leisure trips are not restricted to nearby regions but to whole countries and continents.

Like many other technologies, from the very beginning automobiles carried an undesirable dark side: traffic accidents. The first death by a motor vehicle was registered in Ireland on August 31st, 1869 [90]. Although the number of fatalities was low at the beginning, fatalities exponentially grew throughout the years given the popularization of cars. One and a half centuries later, road accidents represent the ninth cause of death worldwide, and the ONU predicts that in 2030 it will be the fifth [233]. Every year almost 1.2 million people are killed in traffic crashes, while the number of injuries rises to 50 million. Furthermore, attending to the increasing automobile productions in low and middle-income countries, these numbers are expected to rise considerably. Figure 1.1 illustrates the number of deaths per 100,000 population as a result of a traffic accident. As can be seen, high income countries and regions such as the United States or the European Union tend to have a lower number of fatalities than the rest, even though the number of vehicles in these countries is high. Before having a look at this map, one could have the wrong idea of thinking that a lower number of vehicles would be reflected in a lower number of traffic accidents and deaths. In the United States there are 0.8 vehicles per capita [66] (the USA is the most motorized country in the world). In Egypt this number is exactly the half [233], however the number of deaths per 100,000 inhabitants is three times the deaths in the USA. Nigeria, with only 0.3 vehicles per capita, has twice the number of deaths per 100,000 than the USA. In fact, it is the longstanding traffic regulation together with the consciousness-raising of this problem which have progressively decreased the number of fatal accidents in high-income countries. On the contrary, low-income

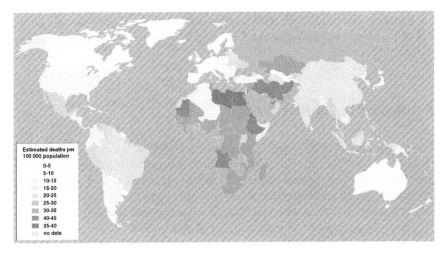

Fig. 1.1 Statistics of deaths related to traffic accidents in the world. It can be clearly seen that although rich countries have higher number of vehicles than low-income countries, the number of fatalities is lower thanks to the improved safety measures, government campaigns and regulation

countries tend to have a higher number of deaths given the opposite reason: lack of regulation in many aspects and low consciousness of the problem. Attending to this, as low-income countries evolve and start increasing their number of vehicles and transport networks the relevance of the problem arises.

According to the International Organization of Motor Vehicle Manufacturers [228], every year around 59 million passenger cars and 20 million commercial vehicles are produced worldwide. This represents an increase of 44 % and 17 % with respect to year 2,000 production, respectively. Although this rate will probably not sustain in very industrialized countries as a result of the increasing price of oil, the popularization of air travel and the *green* trends, it will be largely compensated by emerging economies such as China or India, with increases of 800 % and 400 % in total number of produced vehicles yearly from 2000 to 2010.

Figure 1.2 illustrates another dramatic fact related to the evolution of two emerging economies. On the one hand the United States and the European Union have average population and a big fleet of vehicles. On the other hand growing countries such as China or India have the biggest populations in the world but an average number of vehicles (between 0.2 and 0.5 per capita) [233]. As previously mentioned, having a smaller fleet does not lead to a lower number of accidents, as it is clearly appreciated in this figure. However, it is clear that as emerging countries increase their number of vehicles, the number of deaths will also rise if no solution is implemented. The most direct solutions to fix this problem are well-known given that they have been developed for decades: researching new technology to increase the safety of vehicles and infrastructures.

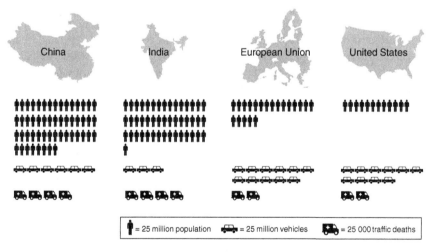

Fig. 1.2 Population, vehicle and traffic deaths statistics of two emerging economies (China and India) and two high-income ones (USA and European Union)

1.2 Advanced Driver Assistance Systems

From the beginning, the automobile industry has proposed new technologies aimed at improving safety and reduce the number of accidents. This technology has progressively gained complexity and improved performance through the years. The first electric headlamps were introduced in 1898 in the Columbia Electric Car, while turn signal lights were devised in 1907. In 1955 Ford included seat belts as an optional equipment and Saab made them standard in 1958, albeit the first proposals and tests of seat belts were made in the 1920s. Volvo presented the current standard three-points belt in 1959, which prevents the vehicle occupants to hit against the interior elements after an impact. The airbag was developed in the late 1960s, however, similarly as with the seat belts it took from two to three decades to become an standard in the United States and in Europe. Nowadays, it is estimated that around 12,000 lives are saved yearly by seat belts and 2,000 by airbags just in the United States [281]. Antilock braking system (ABS) and electronic stability control (ESC) were introduced by Bosch and Mercedes-Benz in 1978 and 1995, respectively. ABS is designed to avoid tires blocking in order to keep their grip constant. ESC is designed to avoid skidding by applying brakes to individual wheels. Both approaches were feasible thanks to the invention of electronics, which provided the necessary fast physics computations required. As can be seen, until the 1990s the technological advances in security relied mostly in physical devices focused on providing safety to the vehicle passengers when the accidents or dangerous situations were happening.

In the last twenty years, advanced driver assistance systems (ADAS) have aimed at predicting dangerous situations and anticipating accidents. These intelligent systems provide warnings and assist drivers to take decisions and even take automatic evasive actions in extreme cases. They are different from the previous technologies in the sense that they do not only rely on physical/mechanical cues from the host vehicle, but they can also *understand* the exterior world or the driver state. In this case, Artificial Intelligence plays a key role when pursuing this understanding. In addition, research in sensors, human machine interfaces and even psychology is also important for these systems.

The pioneer in ADAS was Ernst Dickmanns' group, from the Universitat der Bundeswehr München (Germany), who in 1986 presented an autonomous driving system able to drive through a controlled scenario (a closed highway) at up to 96 km/h [67]. The system consisted of cameras, rudimentary image processors and Kalman filtering. This research later lead to the first European project on autonomous vehicles, called Prometheus.

Nowadays many different ADAS may be found in the marked. For example, the adaptive cruise control (ACC) keeps a constant distance to the front vehicle by slowing down or accelerating the host vehicle. It was introduced by Mercedes and Jaguar in the late 1990s [157]. Lane departure warning (LDW) systems warn the driver when the vehicle moves out of its lane, unless the corresponding direction turn sign is on. It was introduced in trucks in 2000 and later in sedans [158]. This technology is being improved by assisting the steering action or warning/intervening in lane changing in case of danger. Finally, one of the currently hot research topics are advanced front lighting (AFL) systems, which control the headlight parameters so that the beam is optimized for different conditions such as driving speed and direction. The reader can refer to different publications that include comprehensive reviews on these systems [31, 311].

1.3 Pedestrian Protection Systems

Traditionally, the technological improvements in automobiles have been addressed to protect the vehicle occupants in vehicle-to-vehicle crashes. On the contrary, in terms of the automobile industry, road users such as pedestrians and bicyclists have not received the same attention. Having a look at the statistics it can be seen that the proportion of pedestrians killed in accidents is considerably high. In 2003 almost 150,000 injured and 7,000 killed pedestrians were reported in the European Union roads [78], representing the second source of injuries and fatalities just after four-wheeled vehicle passengers. The United States' numbers are similar, counting 70,000 injured and 4,000 killed [281]. In low and middle income countries this number can neither be neglected. As an example, just Delhi City (India) registered almost 1,000 killed pedestrians in 1994 [217], and at the moment this number is likely to have grown given that the number vehicles has been doubled in the country [280]. Statistics also state that 70 % of the people involved in car-to-pedestrian accidents were in front of the vehicle, of which 90 % were moving [137].

Figure 1.3 illustrates the percentage of pedestrian deaths over all the traffic deaths (vehicle passengers, motorbike passengers, cyclists, etc.). A relation can be seen, with some exceptions, between the percentage of pedestrian fatalities and the countries income. For example, in low income countries such as Dem. Rep. of Congo or Kenya more than 50 % of traffic deaths are pedestrians, while this statistic decreases to 10–20 % in high-income countries/regions such as the United States or European

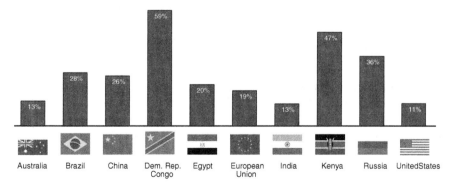

Fig. 1.3 Percentage of pedestrian deaths over all the traffic deaths in several regions of the world

Union. In the future, as the number of vehicles rises in low-income countries, the percentage of vehicle passengers will increase (as a result of having more drivers and occupants) with respect to the pedestrians.

In view of these terrible statistics, during the last twenty years automative companies and suppliers have progressively turned their safety efforts also to pedestrian protection. At the beginning, research was focused on optimizing the vehicle's physical parts in order to minimize impact severity. This research direction is often referred to as improving safety through design. Some examples are collapsing fenders, hood and windshield, or increasing the space between (a softer) hood and the engine to accommodate the pedestrian's head in the case of a crash. Later, more intelligent approaches were introduced. The first investigations on intelligent systems addressing pedestrian protection were conducted in the 1990s by Papageorgiou (MIT), Gavrila (Daimler), Broggi and Bertozzi (University of Parma). Nowadays, pedestrian safety has become an interesting research and development topic for companies, governments and research centers. Some examples of such interest can be seen in the last three European Union Programmes for research, also known as Framework Programmes (FP):

- Under 5th FP: PROTECTOR (4.4 million €, 2002–2003) and SAVE-U [268] (8 million €, until 2005), with Faurecia as coordinator, and CEA, Volkswagen AG, Daimler Chrysler AG, Siemens VDO Automotive AG and Mira Ltd as partners.
- Under 6th FP: PReVENT's APALACI [256] (3.75 million €, 2004–2007), coordinated by FIAT with partnering from Daimler Chrysler AG, Robert Bosch GmbH, Ibeo Automobile Sensor GmbH, Volvo and University of Parma.
- Under 7th FP: ADOSE (10.2 million €, 2008–2011), coordinated by FIAT; and FNIR [96] (3.12 million €, 2008–2010) coordinated by Autoliv AB.

Pedestrian Protection Systems (PPSs) are a particular type of ADAS focused to pedestrian safety. A PPS is formally defined as a system that detects both static and moving people in the vehicle's surroundings (typically in the front area) in order to provide information to the driver and perform evasive or braking actions on the host vehicle if needed. Pedestrian detection before the impact (either long or short term) is crucial given that the severity of injuries for the pedestrian decreases with the crashing vehicle's speed [11]. Hence, any reduction in the speed can drastically reduce the crash severity. According to [11], pedestrians have a 90 % chance of surviving to car crashes at 30 km/h or below, but less than 50 % chance of surviving to impacts at 45 km/h or above. The benefits of PPSs are twofold: (1) PPSs are able to minimize the reaction time with respect to the human drivers (in humans it depends on age, hours of continuous driving, alcohol consumption, distractors, day/night, etc.) and (2) they can control active measures such as airbags or brakes to minimize the potential impact.

1.4 The Role of Computer Vision

The central problem of PPSs corresponds to the task of detecting pedestrians. In order to detect objects (e.g., vehicles, pedestrians, obstacles) in the distance, ADAS use sensors that provide data to a computer/controller that processes them and performs the corresponding actions. The most widely used sensors for pedestrian detection are cameras mainly thanks to the rich high resolution information they provide, with cues such as edges, contours, texture or even relative temperature. In order to capture these cues, cameras working either in the visible or infrared spectrum are used.

A pedestrian detector must tackle several challenges, which are independent from the used sensor:

- A high variability in pedestrians' appearance since they can be of different size (especially in height), change their pose, wear different clothes and carry different objects. Moreover, pedestrians appear at different viewing angles (e.g., lateral and front/rear positions) and are imaged from a large range of distances (at least 25 m and up to 60 m, which corresponds to 60–15 pixel pedestrians, depending on the image acquisition system.
- Deal with cluttered background under a wide range of illumination and weather conditions, and in the presence of shadows and poor contrast. Furthermore, pedestrians can be partially occluded by vehicles, street furniture or other pedestrians.
- Manage highly dynamic scenes where not only pedestrians and other objects are in motion but also the camera is moving.
- Different temperatures and distances affect nighttime detection with infrared cameras.
- A high demanding performance in terms of accuracy (false alarms and misdetections) and system reaction time.

As can be appreciated, the topic differs from general human detection systems such as surveillance or human-machine interfaces. Although PPSs can indeed make use of the techniques developed for these applications, many typical simplifications as the following must be discarded attending to the inherent challenges of PPSs:

- Static camera assumption, common in surveillance, is not applicable. Hence, traditional techniques related to background subtraction cannot be directly applied in this research.
- Indoor illumination, common in human-machine interfaces, does not suit driving assistance applications.
- Model size is typically more constrained in dataset retrieval than in PPSs. For example, the human model in visual dataset searching is normally focused on well-seen people with a considerable amount of pixels to analyze. In the case of PPSs, pedestrians at 50 m can measure up to 10 pixels high. In the case of retrieval, however, the pose variability is more flexible than in PPSs, in which pedestrians are assumed to stand up on the road or pavement except in very rare cases or in the case of children [14].

The reader can refer to the surveys in [97] and [215], focused on generic human detection, and [187], focused on intelligent transportation systems, for more details about human detection for applications different from PPSs.

1.5 Generic Framework

A basic pedestrian detector is often based on two components: one that selects image regions likely to contain a pedestrian and another that classifies the regions as pedestrian or non-pedestrian. These two parts are usually called candidates generation and classification. Many object detection algorithms such as face or vehicle detection use this two-step approach. As it will be seen in further chapters, the development of new techniques not only has led to the improvements of these two components but also to the inclusion of additional ones. For example, since a PPS is focused on the continuous detection of pedestrians, it seems natural to include a new component capable of *following* the pedestrians through time. This would provide information on the direction of the targets, for instance. From here on, we will refer to these components as modules. Breaking down the systems into several modules is a convenient way to favor its understanding. For example, in Sun's vehicle detection review [286], techniques are divided into hypothesis generation and hypothesis validation, thus allowing the reader to concentrate on the methods for solving simpler problems, rather than approaching the problem as a whole.

It is not easy to divide the PPSs proposed in the literature in specific modules: sometimes there is not a clear description of them, sometimes different modules are mixed together in a single algorithm, and even the functionality of a module is completely omitted. In order to provide an organized and comprehensive review of the existing techniques, we first describe a generic architecture of six different modules. Each module has its own objectives and responsibilities in the system, so by fitting each algorithm in the literature in its corresponding module it will be easier to compare and analyze the proposals. The used modules are the following:

- Preprocessing: it takes input data and prepares it to further processing. The data comes mainly from a camera but in some cases information is also acquired from car sensors, odometers, etc. The tasks carried out are diverse, some examples are sensors synchronization, adjust camera exposure time and gain, and calibration.
- Candidates Generation: it extracts regions of interest (candidates) from the image to be sent to the classification module avoiding as many *non-pedestrian* regions as possible.
- Classification: it receives a list of candidates to be classified as pedestrian or non-pedestrian.
- Verification and refinement: it verifies and refines the candidates classified as pedestrians, referred to as detections. The verification filters false positives using criteria not overlapped with the classifier while the refinement performs a fine

segmentation of the pedestrian (not necessarily silhouette-oriented) so to provide an accurate distance estimation or to support the following module, tracking.

- Tracking: it follows the detected pedestrians along time with several purposes such as avoiding spurious false detections, predict the next pedestrian position and direction and even other high-level tasks such as inferring pedestrian behavior.
- Application: it takes high level decisions (braking, steering, etc.) by making use of the information provided by the previous modules. This module represents a complete area of research, which includes not only driver monitoring but also psychological issues, human-machine-interaction, etc.

Figure 1.4 shows a schematic view of the PPS architecture. It is worth noting that although no feedback arrows have been included, several works make use of feedback or iterative cooperation between modules, e.g., classification and candidates generation.

An important aspect of the book is that it is focused on works using passive sensors, cameras working either in the visible (typically for daytime) or infrared (for nighttime) spectra, which are the most commonly used sensors for PPSs. Henceforth, we will refer to the visible spectrum as VS (i.e., the range 0.4–0.75 μm) and to the infrared either as NIR (near infrared, 0.75–1.4 μm) or FIR (far infrared, 6–15 μm).[1] The sensibility of NIR sensors ranges from 0.4 to 1.4 μm, so it can be said that they work in the VS+NIR spectrum. Regarding FIR sensors, they capture relative temperature, which is very convenient for distinguishing targets such as pedestrians or vehicles from asphalt or trees. For an analysis of radar, laserscanner or ultrasonic sensors please refer to [50].

1.6 Book Outline

This book surveys the state of the art in pedestrian detection for PPSs putting the emphasis on two aspects: (1) to overview the techniques used in PPSs from the first of these systems to the latest ones; (2) to provide a global viewpoint not specifically focused on classification techniques but also on the different ingredients of a complete PPS. The book can be seen as an extended version of the survey in [121] with updated references and comments. Note that in this survey we have deliberately avoided a quantitative performance assessment between the existing techniques, as this is out of the scope of the book. For surveys specifically evaluating this aspect for some selected proposals the reader can refer to the excellent surveys by Hussein et al. [152], Dollár et al. [74] and Enzweiler et al. [83]. Additionally, for a more generic survey on the pedestrian safety measures from a transportation viewpoint (not only on-board PPS but also infrastructures) the authors propose [109] as a the relevant publication to read.

---

[1] While it would be more precise to refer to this range as long wave or thermal infrared, in this book we use the term FIR (which in fact corresponds to the 15–1,000 μm range) given that this is the most common naming in the pedestrian detection literature.

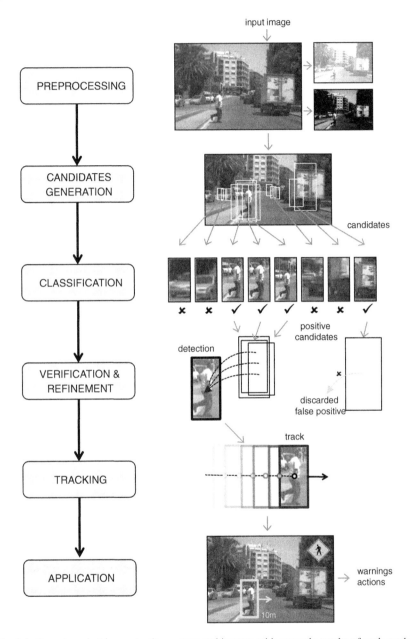

Fig. 1.4 Generic pedestrian protection system architecture with example results of each module. Note that in some proposals several modules are combined in a single one or can loop back the results between them, we have not included these case for the sake of clarity

The following two chapters are focused on the two core modules of the system: candidates generation in Chap. 2 and classification in Chap. 3, this latter being the longest of the book as it concentrates most of the literature. The rest of the system components (including preprocessing, verification and refinement, tracking and application), together with an overview of the aspects related to real-time, are surveyed in Chap. 4. Chapter 5 analyzes the existing methodologies for evaluating both classification and complete systems, including protocols and databases. Finally, in Chap. 6 we present the overall conclusions and the general perspective for the future of this research area.

Chapter 2
Candidates Generation

Candidates Generation extracts regions of interest (ROIs), usually rectangular windows, from the input image that are likely to contain a pedestrian avoiding as many non-pedestrian regions as possible. These ROIs will be sent to the classification module. Candidates generation is also referred to as foreground segmentation [125] or candidates extraction [4], though the most common name is candidates generation. It is the first core module of a PPS, and the approaches can be divided into 2D-based, 3D-based and motion-based.

2.1 2D-Based Approaches

The simplest and most extended candidates generation approach is the sliding windows algorithm. It is used in many object detection systems, especially in the ones in which there is barely any constraint on the candidates' position and size. It consists in exhaustively scanning the input image with window candidates only constrained by aspect ratio and perhaps position (e.g., avoiding the image's upper-part) but not by any complex reasoning (Fig. 2.1).

One of the first approaches using sliding windows is presented by Papageorgiou et al. [241]. The authors propose to scan the input image with windows of 128×64 pixels to be sent to the classifier. In order to achieve multi-scale detection, the authors propose to scan the image from 0.2 to 2.0 times its original size with increments of 0.1. This candidates generation procedure was popularized after the comprehensive study by Dalal in his PhD Thesis [59], but it is successfully used in many human detection systems, e.g., surveillance, indexing, sports video analysis, etc. There exist two different approaches to multi-scale sliding windows depending on what element is scaled: the image or the detection window. The former approach, which is used in [59, 83, 93, 189, 241, 299, 328], consists in an image-pyramid of s scales (typically from 8 [60] up to 50 [72]) that is scanned with a constant-size detection window. In the second approach the detection window is resized in s different

D. Gerónimo and A. M. López, *Vision-based Pedestrian Protection Systems for Intelligent Vehicles*, SpringerBriefs in Computer Science, DOI: 10.1007/978-1-4614-7987-1_2, © The Author(s) 2014

Fig. 2.1 Sliding windows concept. A windows scans the input image in all the possible positions. In real systems, the resulting windows overlap and are of different sizes

scales to scan a fixed size input image [123, 245, 310]. Notice that the type of scanning limits the use of one feature or another, i.e., the local image features in a resizing detection window must be also resizeable, while in the image-pyramid scanning they are not restricted. These aspects, together with a description of intermediate scanning methods aimed at accelerating the detection process, will be detailed in Chap. 3.

Although sliding windows is successfully used in many different computer vision applications, in the case of PPSs this technique has two main drawbacks. First, the number of candidates is large, which makes it difficult to fulfill the real-time requirements. As an example, a regular scan over a 640×480 pixels image can provide from hundreds of thousands to millions of windows, depending on the sampling step and the minimum window size. An intuitive procedure is to decrease the sampling step to reduce the number of windows, however this also decreases the chance that a candidate window correctly frames the pedestrian in an unbiased manner for its further classification. Second, many irrelevant regions such as sky regions or windows inconsistent with perspective are likely to be sent to the classification, which increases the potential number of the system's false positives. An intuitive approach to reduce this kind of candidates is to discard the top 1/3 of the image. However, this solution does not avoid perspective-incorrect candidates such as small-sized windows in near road positions. A more sophisticated approach that depends on the classifier used but it is applicable in any context is proposed by Lampert et al. [174]. This proposal makes use of a branch-and-bound technique that bounds the classifier output, providing a globally optimal solution at sublinear time. The authors successfully implement the proposal using different classifiers, e.g., SVM using spatial pyramid kernels or χ^2-distance. However, as it will be seen later, there are more clever approaches that take advantage of the application's information.

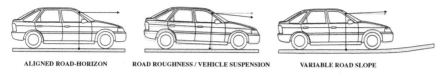

ALIGNED ROAD-HORIZON   ROAD ROUGHNESS / VEHICLE SUSPENSION   VARIABLE ROAD SLOPE

Fig. 2.2 Different situations in which the relative camera–road position varies

Fig. 2.3 Flat world assumption concept. The *horizon line* computed as the road position at infinity is one of the ways to represent the camera–road position

Pedestrian detection systems soon started to include some scene's prior knowledge aimed at improving sliding windows. The approach called *flat world assumption* defines the techniques that restrict the candidate search to the ground plane without any other information than the initial camera position with respect to the plane. It consists in fixing a constant camera-road position, represented as a fixed 3D plane or a horizon-line position, and then scanning the road with pedestrian-sized windows (see *Aligned Road-Horizon* in Fig. 2.2). Accordingly, the candidate windows will be projected in specific places of the image depending on their position and distance (see Fig. 2.3). By sampling just the estimated road the number of candidates is reduced to tens of thousands of windows, and again this number depends on the scan's density and the maximum detection distance. Different research groups, both in pedestrian [23, 113] and vehicle [254] detection made use of this idea. This technique performs significantly well when dealing with non-urban roads such as highways, where in general the road slope does not change much and in which the vehicle dynamics (e.g., braking, accelerating) are smooth. In order to solve the problem of changing camera-road position, Ponsa et al. [254] fix three different scanning road planes to detect vehicles in highways. Nevertheless, this is insufficient for pedestrian detection in urban scenes, since not only the target is smaller but also the road slope changes a lot in cities and the vehicle dynamics affect the angle between camera and road. Accordingly, some researchers tried to amend this problem by using visual features. In [22, 44], Broggi et al. correct the vertical image position by relying on the detection of horizontal edges oscillations: the horizon line is computed according to the previous frames. A comparative study of different monocular camera pose estimation approaches has been presented in [32]. It includes horizontal edges, features-based and frame difference algorithms. The following step related to this approach goes through the use of 3D information to estimate the relative road-camera position, which is described in next section.

Other non-related 2D-based approaches use the image information to generate the candidate windows. For example, Miau et al. [156, 206] use a biologically inspired attentional algorithm that selects windows according to pixels' color, intensity and gradient orientation. In several works from Parma University, the vertical symmetries in the visible [23, 26, 39, 41] and FIR spectra are used alone [22, 25] or as a complement to stereo imaging [39]. In this case, windows are adjusted around each symmetry axis maintaining the pedestrians' proportions. The presence of many horizontal edges is often taken as a non-pedestrian quality. Hoiem et al. [142] presented a probabilistic framework for 3D geometry estimation based on a monocular system. A training process based on 60 manually labeled images is applied to form a prior of the horizon position and camera height (i.e., camera pose values). In the same vein, scene understanding approaches such as [3, 127, 282] can help the generation process by making use of statistical patterns of the scene. This statistical process, which can be seen as a pre-classification step, is interesting since higher level information is used, always that the computational time is lower than a traditional candidates generation and classification sequence. The well-known *implicit shape model* (ISM) classification technique (Chap. 3) can also be seen as candidate generation. In this technique, the image keypoints (Harris, Hessian, etc.) matching the ones in a pedestrian model vote to promising locations in the image likely to contain pedestrians.

A recent trend is to reduce the number of candidates by performing a sliding windows-like coarse-to-fine search. Pedersoli et al. [246] perform a multi-resolution cascade-based search that either prunes or evaluates candidates based on the confidence of a regular classifier at a given sliding windows scale. Gualdi et al. [131] use a particle-based approach to estimate the density function of detections and then draw particles in the regions where the classifier confidence is higher. Dollár et al. [69] propose to take advantage of the correlated classifier responses of adjacent candidates by exciting or inhibiting neighbor windows, providing 4–30 times speedups.

Intensity thresholding is the simplest segmentation technique used in NIR/FIR imagery. The main methods are single thresholding [283], double image and hot spots-based thresholding [43], adaptive intensity-based thresholding [291, 297], FIR models [42] and vertical and horizontal histogram projection with thresholding [28, 91]. *Hypermutation networks* (HN) [227], which use a multi-stage neighborhood pixel classification, are considered in more sophisticated approaches such as [196, 204] to classify pixels as foreground/background. In the case of [196], HN are combined with Chamfer System (Chap. 3) and flat world assumption. In [204], output pixels from the network are grouped using connected component analysis so the algorithm may be understood as a segmentation/classification process. Sun et al. [285] propose to filter the sliding windows candidates not close to the neighborhood of detected SUSAN [279] keypoints.

2.2 3D-Based Approaches

One of the first stereo algorithms specifically developed for ADAS is presented by Franke et al. [102]. The method performs stereo correspondence by classifying local structure to resolve the ambiguities in the disparity histogram. The algorithm is later extended with a multi-resolution method with sub-pixel accuracy in [99, 101]. The output of these methods is used in the well-known PROTECTOR system [113]. In this system, the stereo map is multiplexed into different ranges and scanned with pedestrian-sized windows. If the depth features in one of the windows exceeds a given percentage, then the window is sent to the classifier. The system uses flat world assumption to reduce the number of candidates while assuming a fixed camera-road position. Zhao and Thorpe [341] use a much simpler approach: they detect discontinuities directly in the disparity image to extract candidates, which are then filtered according to size and shape constraints.

Later on, stereo has been demonstrated to be a very reliable cue to compute an accurate camera-road pose estimation in cases in which it is not constant (see center and right illustrations in Fig. 2.2). In [175], Labayrade et al. introduced the first frame-based road-camera pose estimation algorithm using stereo: the v-disparity representation. This representation is based on the fact that a plane (e.g., the road surface) in Euclidean space becomes a straight line in the v-disparity space. It consists in accumulating stereo disparity along the image y-axis in order to (1) compute the road's slope, which is related to the horizon line and (2) point out the existence of vertical objects when the accumulated disparity of an image row is very different from its neighbors. Extensions to this representation may be found in [145]. Figure 2.4 illustrates the v-disparity image next to two input images. The diagonal line represents the road's position while the vertical bars highlight the existing vertical objects, e.g., pedestrians. As can be seen, the scenario in the right image is more challenging as the street is narrow and pedestrians are not isolated, which produces a noisier v-disparity map that makes the pedestrians extraction more difficult. Alonso et al. [245] point out this difficulty of using v-disparity in cluttered scenarios, and propose to use an alternative approach for their candidates generation module. First, they apply subtractive clustering to the 3D points of the scene making use of the points' density. Then, starting from the corresponding 2D window, a set of oversized, downsized and shifted windows are generated to get around 15 windows for each original cluster.

Other road pose estimation approaches work directly in Euclidean space. For instance, Sappa et al. [267] propose to fit 3D road data points to a plane (Fig. 2.5), whereas Nedevschi et al. [226] use a clothoid. In the Euclidean space classical least squares fitting approaches can be followed while in the v-disparity space voting schemes are generally preferred (e.g., Hough transform). Ess et al. [88, 89] use

Fig. 2.4 v-disparity representation (showing only the *right* image and the computed v-disparity). Each input image's pixel votes depending on its row and disparity to the v-disparity image

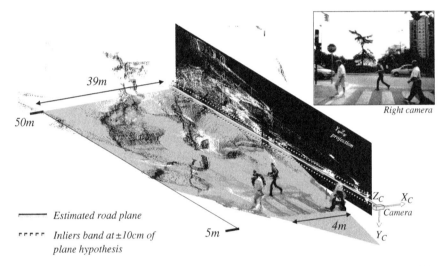

Fig. 2.5 Road fitting representation: the scene's 3D points are used to estimate the road position by using RANSAC [267]

pedestrian location hypotheses together with depth cues to estimate the ground plane, which is used to reinforce new detections-tracks. The authors call this approach cognitive feedback in the sense that a loop is established between the classification and tracking modules and ground plane estimation. The same idea, but using detected vehicles width to estimate the ground position, had been proposed before in [254] by Ponsa et al. In [166], Keller et al. propose to track the estimated road plane (which is represented by a B-Spline) with a Kalman Filter. Attending to all these works, one clear conclusion is that the road position is a crucial information to be estimated (although there are works that do not use it [225]). Figure 2.6 illustrates the detection results in several consecutive frames when using road estimation (after classification and refinement modules).

Fig. 2.6 Pedestrian detection using ground plane position estimation when driving through a speed bump [119]. The camera suddenly moves downward and the *horizon line* as well as the objects in the scene move upward

Sudowe et al. [284] filter the output of a traditional sliding windows algorithm by geometrical constraints taking into account the estimated ground plane. In this case, instead of generating ad hoc windows for the current road plane, they propose to use the original sliding windows ones, hence a HOG-like pyramid image scale (see Chap. 3) can be used.

Researchers from Daimler AG have recently developed a new technique to represent vertical objects in the scene: the *stixels*. The technique is originally proposed by Badino et al. [13] and it is likely to be a key component in future PPSs given its simplicity and potential in the candidates generation module for any type of target. The algorithm first computes the occupancy grid from 3D data, in which the evidences of scene occupancy are represented in a matrix placed over the ground. A B-Spline is fit to the 3D data to adjust the plane surface instead of assuming a planar road. Then, dynamic programming is employed to detect the points where the free road and the bottom of each obstacle is located. The object's height is computed in the same way. The output of this technique is a set of vertical poles of, e.g., 20 cm wide that represent vertical objects placed on the road (stixels). Figure 2.7 illustrates the idea.

Several advances have been proposed based on this concept. For example, Pfeiffer et al. [248] propose to track stixels over time by using the so-called 6D-vision

Fig. 2.7 Stixels representation of a typical urban scene (*red* and *green* stixels represent close and far *vertical* objects, respectively) [249]. Image courtesy of David Pfeiffer (Daimler AG)

tracking [103], aimed at grouping stixels with similar motion, which can be used to generate candidates. The same authors have embedded the stixels computation in a probabilistic framework that is capable of assigning different depths to each stixels column (called multi-layered approach) [249], which can also help to generate candidates by grouping stixels at the same depth. The potential of the stixel representation as a candidates generation technique has been proven in vehicle and pedestrian detectors such as [17, 18, 86, 247].

As a final remark regarding 3D-based algorithms, we refer the reader to the recent survey in [192] to have an extended review of these approaches.

2.3 Motion-Based Approaches

Inter-frame motion and optical flow [79] have been used for candidates generation primarily in the general context of moving obstacle detection. Franke et al. [100] propose to merge stereo processing, which extracts depth information without time correlation, and motion analysis, which is able to detect small gray value changes in order to allow early detection of moving objects. In [180], Leibe et al. present a real-time Structure-from-Motion based approach for ground plane estimation. This online estimated plane is used to update the camera calibration and thus to segment objects from the ground surface. In [87], Enzweiler et al. make use of learned motion parallax features (horizontal velocity and local parallax flow) as candidates generator. The authors point out that with parallax flow it is possible both to detect static and moving pedestrians.

One of the current promising motion-based techniques to generate candidates is the 6D-vision algorithm [103]. This technique tracks specific 3D points in consecutive images by fusing depth and motion information with Kalman Filters. Since it is capable of tracking objects at a pixel-level, it has the potential of discriminating static from moving objects in the scene [261].

2.4 Discussion

3D-based systems present advantages such as the good accuracy and robustness in different conditions. The resolution of the stereo reconstruction is sufficient to successfully detect pedestrians in the range from 5 to 30 m. Since the stereo information can be used in different PPS modules (i.e., not only candidates generation but also classification [81, 85] or tracking) its computation time is well-spent. The drawbacks of stereo systems mainly come from the errors in the reconstruction owing to non-textured regions, calibration methods, etc. However, in the last years very significant advances have been made in the field: parallel stereo reconstruction [302], large-scale stereo matching [118], global optimization methods [45, 49], etc. We refer to [269] for a survey study on two-frame stereo correspondence algorithms and their website[1] for an updated algorithms evaluation.

Multi-modal stereo analysis (i.e., stereo computation from visible and infrared sensors) is one of the interesting topics currently researched to generate candidates. For example, systems such as [15, 170, 171] extract stereo information from a visual spectrum camera and an infrared camera. This approach corresponds to sensor fusion. Active sensors such as laser scanners have their role in the short-time collision detection given that they are fast and very accurate at short distances. It also worths mentioning time-of-flight sensors, which pose as an alternative laser scanners [317].

Finally, motion and 2D-based approaches, specially coarse-to-fine search methods, can be a great complement to reduce the number of candidates produced by 3D cues at a very low computational cost.

---

[1] http://vision.middlebury.edu/stereo

Chapter 3
Classification

The classification module receives a list of *candidates* (rectangular windows in practice) that are likely to contain a pedestrian. In this stage, such candidates are classified as *pedestrian* or *non-pedestrian* (Fig. 1.4) with the goal of minimizing the number of wrong decisions while maximizing right ones. Broadly speaking, this is a pattern recognition module involving vision and machine learning. The former field dealing with image descriptors and pedestrian models. The later one providing algorithms to learn, mostly from labeled samples, the pedestrian/non-pedestrian decision rule (i.e., the pedestrian classifier) based on the mentioned descriptors and models.

Pedestrian classification is indeed a very hard pattern recognition problem due to the circumstances enumerated in Sect. 1.4. Briefly, such circumstances challenge classifiers to be robust to (Figs. 3.1–3.4):

- changing outdoor conditions: background and illumination/temperature;
- different inherent pedestrians appearance: clothing, size and pose;
- imaging principles affecting: viewed side (owing to the relative position between the camera and the pedestrians), perceived details and degree of focus (owing to the distance from the camera to the pedestrians);
- the presence of other traffic participants (urban furniture, vehicles and other pedestrians) that cause partial occlusions.

From the PPSs' viewpoint the ideal approach would be to design a classification module that progressively simplifies the pattern recognition problem. For example, first removing outdoor variability, then normalizing pedestrians appearance, etc. and finally reducing it to a sort of correlation with a simple pedestrian canonical model. Unfortunately, there does not exist such a straightforward solution. Instead, the robustness to the combination of the exposed variability sources must come from a joint action of the image descriptors, the pedestrian pictorial model as well as the chosen pedestrian classifier type together with its training.

Nevertheless, some of such components are expected to provide a certain degree of invariance with respect to specific variability sources. For instance, some image descriptors are designed to provide as much robustness as possible to cope with different global scene illuminations, presence of shadows, etc. Parts-based pedestrian

D. Gerónimo and A. M. López, *Vision-based Pedestrian Protection Systems for Intelligent Vehicles*, SpringerBriefs in Computer Science, DOI: 10.1007/978-1-4614-7987-1_3, © The Author(s) 2014

Fig. 3.1 *Left* different scene illumination. *Right* partial occlusion versus full visibility example

Fig. 3.2 Changing pose under the same side view

Fig. 3.3 Changing resolution (and *background*) due to distance to the camera (*motion*)

models based on such image descriptors are intended to add robustness to large pose changes. Multiple view ensembles of parts-based models try to gain robustness to imaging view point, and so on. Accordingly, in the rest of this chapter we review the most successful ideas present in the literature regarding image descriptors, pedestrian models, and the construction of pedestrian classifiers.

3.1 Preliminary Concepts

3.1.1 Image Descriptors

Vision-based PPSs rely on a specific image *modality* or a combination of them. These modalities are *temperature, luminance/color, depth* and *motion flow* (Fig. 4.2). From the point of view of a candidate window each modality provides per-pixel raw information (assuming dense depth and flow computation). Therefore, such raw

Fig. 3.4 Frames taken from a sequence of just 15 s. Note the huge variability in clothes, poses, sizes (resolution), and views of the pedestrians

information can be processed to obtain *descriptors* that enhance *cues* of interest (e.g., luminance edges) and is robust to some of the adversities we have enumerated before (e.g., global illumination changes in the luminance modality). Since these descriptors will be later the raw information employed to learn/apply the desired pedestrian classifier, we can also term them as *features* following the machine learning terminology.

Most of the works in the literature focus on the luminance modality, thus, along the rest of the chapter we will assume such a modality unless another one is explicitly mentioned. However, we remark that many descriptors can be applied to different modalities, i.e., they are not modality dependent. In some cases, the input image is in RGB format and the descriptor operates with it by following a *max* style. For instance, computing per-channel image gradients at each pixel but only considering those with the maximum magnitude for further pixel description. We consider such scheme as a type of luminance modality since there is not an explicit description of color.

3.1.2 Pedestrian Models

Along the years models of increasing complexity have been developed to distinguish pedestrians from background. The one termed as *holistic* considers pedestrians as a whole, i.e., with no parts. In contrast, for the so-called *parts-based* methods, pedestrians consist of a set of body-inspired parts, anatomy-based (e.g., head, legs, arms, etc.) or location-based (e.g., bottom, middle, top candidate window sections), often complemented with full body (holistic) as well. Additionally, both holistic and parts-based approaches can explicitly incorporate *multi-view*, *multi-resolution*, and *occlusion* modeling.

3.1.3 Pedestrian Classifiers

As mentioned before, a pedestrian classifier decides whether a candidate window embodies a pedestrian or not. Unless opposite stated, we assume that all candidate windows reaching a particular pedestrian classifier have the same *canonical* size.

Most pedestrian classifiers are obtained through a machine learning process following the *discriminative* paradigm, i.e., learning from labeled samples. Popular learning procedures used in the pedestrian detection context are support vector machines [303] (SVM; e.g., linear, kernelized, latent, etc.), AdaBoost [104] variants (e.g., Real AdaBoost, LogitBoost, GentleBoost, MPLBoost, etc.), different types of neural networks [30] (e.g., multi-layer perceptron, convolutional neural networks, etc.), as well as different combinations of such procedures. As we will see, in all cases the training data and some additional *tricks* are of paramount interest for obtaining better classifiers.

3.2 Holistic Models: Focus on the Features

Classifying the human body as a whole is a very simple approximation in itself, so works following this approach actually focus on searching for the best image descriptors/features and types of classifier. Therefore, we organize the rest of the section according to the core features used for building the holistic pedestrian classifier.

3.2.1 Templates

For most of the features in this section (see later Haar, EOH, HOG, LBP, etc.) pedestrian and background samples are required to obtain a decision boundary from them by using a selected machine learning procedure. This is not the case for *templates*, which are pedestrian descriptors that can be defined using only pedestrian examples, i.e., not requiring counterexamples. For instance, pedestrian silhouette shapes are rather characteristic. Accordingly, we can think in a very simple pattern matching approach to detect pedestrians. Let us assume that we have computed off-line a binary template representing pedestrians by a closed contour-like curve delineating a sort of average silhouette. Then, given an input image (e.g., a luminance image) we

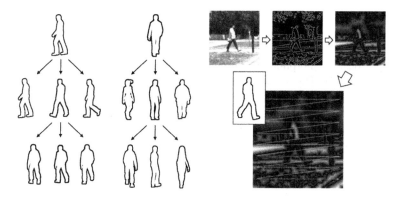

Fig. 3.5 *Left* hierarchy of pedestrian silhouettes (inspired by [110]). *Right* example of template matching based on a silhouette model searched in the distance transform space of the edges of the input image. At each search position average chamfer distance can be used as pattern similarity. Scaling the pattern or the input image can be done for multi-scale search

extract its contours. Now, the pattern matching procedure would consist in *placing* the template at different *positions* of the image of contours. At each of such positions, a *similarity measure* between the template and the image of contours would determine if there is a pedestrian or not, i.e., depending on whether the similarity reaches a supplied threshold or not. Note that background samples are not required for obtaining the pedestrian silhouettes (templates).

Unfortunately, real-world urban scenarios are so cluttered and pedestrians' silhouettes so variable that such a naive procedure would not work. However, robustifying this basic procedure in order to exploit silhouette information is worth to be pursued. Accordingly, in [110, 112, 115], Gavrila and Philomin exploit (binarized) contour information to detect pedestrians. In their approach, there is not a unique pedestrian template silhouette but a hierarchy of them automatically built from example silhouette annotations (Fig. 3.5-Left). This hierarchy allows coarse-to-fine search at the same time that it captures the variability of pedestrian silhouettes in a more fine grained manner. In addition, finding a contour template in a contour image is difficult because only when the template is very close to one of its instances in the image the used similarity measure gives a high response. In other words, the similarity measure offers a sort of desertic landscape with abrupt peaks difficult to find by local search, especially for large search spaces depending on template location, rotation and scaling parameters. To solve this problem, Gavrila and Philomin transform the contour image using the chamfer distance transform [33] (Fig. 3.5-Right). Henceforward, the average chamfer distance was used as similarity measure. This turns out in a smooth similarity landscape where optimized search can succeed, specially the coarse-to-fine search induced by the hierarchy of pedestrian silhouette templates. This approach is known as *chamfer system*.

Broggi et al. also use a pattern matching approach in [39] for detecting pedestrians. However, the aim of this work is more focused on candidates generation and verification. For classification a simple upper-body contour template is correlated to the edge modulus of each candidate window, considering several scales of the template. For FIR images, Bertozzi et al. [22, 27] use a very simple binary mask encoding pedestrian morphology as template, which is correlated to the original FIR candidate windows. This work is further improved in [25, 40], where a 3D synthesized pedestrian with simulated body temperature is used to generate a bunch of 2D templates varying in pose and point of view. However, the generated templates are not sufficiently realistic as to rely on correlation to determine whether a FIR candidate window contains a pedestrian or not [42]. In [222], Nanda and Davis also used probabilistic pattern matching for detecting pedestrians in FIR images.

Leibe et al. introduce the so-called *implicit shape model* (ISM) in [181, 183]. The ISM is built by extracting local appearance features from object examples and modeling their position probability distribution with respect to the object's center. Feature extraction consists in building a *visual codebook*. More specifically, an interest point detector is applied to each example, then local patches around each detected point are stored. All the extracted local patches from all the examples are clustered according to their appearance (agglomerative clustering and normalized correlation is used). The set of clusters' centers is the visual codebook. Then, all examples are revisited to collect the relative positions of the codebook entries (e.g., the examples' bounding box center could be used as a reference position). In particular, each entry can be found at different positions (i.e., where the normalized correlation of the entry and a local patch is larger than the matching threshold used for clustering during codebook building). Therefore, each codebook entry has an associated position density distribution (a non-parametric function is used). The codebook together with such distributions form the ISM.

In order to detect pedestrians in an image an interest point detector is applied first, and then the interest points' associated local patches are collected (Fig. 3.6). All collected patches are matched in the codebook (using normalized correlation) and their associated relative positions are used to vote according to a generalized Hough transform, so that maxima in this space (detected with mean-shift) are supposed to correspond to object centers. Then, the local patches that voted to each maximum are considered. In fact, rather than considering only the maxima-voting patches provided by the interest point detector, a more dense patch sampling around them can be used to obtain object detections including pedestrian-background border. Conflictive detections due to overlapping are resolved by a criterion based on minimal description length [181]. Since interest points are detected at an associated scale, they give the possibility of voting not only for a position but also for a scale, which is interesting to tackle different object sizes. Finally, as aforementioned, ISM allows to infer a probabilistic segmentation of the detected objects. Basically, this is accomplished by storing a segmentation mask per codebook entry and position. Then, the segmentation can be obtained by backprojecting to the image the masks associated to the entries supporting the detections, which is done taking into account the confidence on each matched patch to weight the respective mask.

Seeman et al. [271] and Leibe et al. [182] analyzed best combinations of interest point detectors and local patch descriptors. It seems that for pedestrian detection a

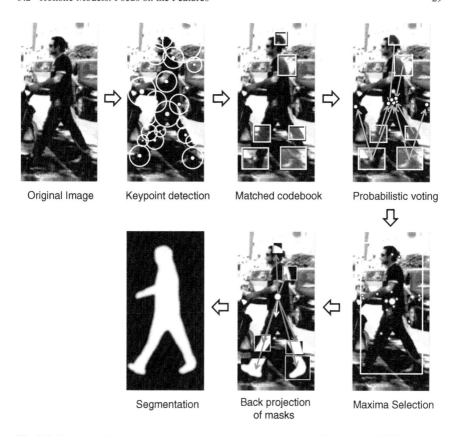

Original Image	Keypoint detection	Matched codebook	Probabilistic voting

Segmentation	Back projection of masks	Maxima Selection

Fig. 3.6 Pedestrian detection and segmentation based on the implicit shape model (ISM)

proper combination consists of Harris-Laplace interest point detector of Mikolajczyk and Schmid [207], and either the scale invariant feature transform (SIFT) by Lowe [194] or *shape context* by Belongie et al. [16] (using the implementation provided by Mikolajczyk and Schmid [208] which includes PCA dimensionality reduction). Note that both SIFT and shape context rely on histograms of gradient orientations (bins at squared cells in the SIFT case, at log-polar cells centered at edge points in shape context). In fact, in [182, 271] the initial ISM proposal of using directly the raw information of local patches to form the codebook (i.e., as explained above according to [181, 183]) showed inferior detection performance than SIFT/shape context descriptors.

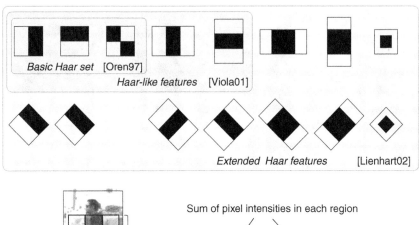

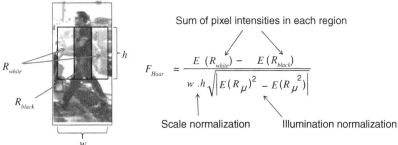

Fig. 3.7 *Top* different sets of filters inspired in Haar wavelets. *Bottom* example of application of a vertical filter

3.2.2 Haar Features

In [232], Oren et al. introduce the basic *Haar features* for pedestrian detection, where the *Haar* name is given due to the similarity of these features with the wavelets introduced by Alfred Haar in 1910 [133]. A basic Haar feature computes the difference of average values between two adjacent rectangular regions of a given image modality. In [232] the regions configuration follow three patterns that roughly emulate vertical, horizontal, and diagonal first order derivatives (Fig. 3.7). The work is extended to face and car detection [240, 241]. In [232] Haar features are separately extracted from the R, G, and B channels of color images, while in [241] only luminance is considered as an speed acceleration step (among others) for pedestrian detection on board a DaimlerChrysler S Class demonstration vehicle.

The different types of basic Haar features are placed within candidate windows by varying their size and location, as well as allowing them to overlap. Therefore, many features can be collected, eventually hundred of thousands. In [232, 240, 241] the features are fed into a SVM with quadratic kernel, which in practice made inviable to consider all them. Accordingly, for pedestrian classification only 29 manually selected Haar features are used.

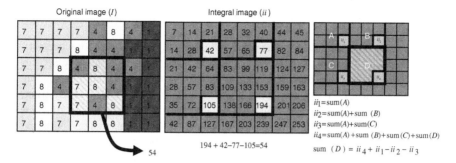

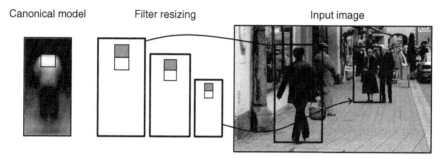

Fig. 3.8 Fast computing of the summation of values within an image rectangular region based on the integral image concept

Fig. 3.9 Haar based filters allow fast filter resizing to accommodate different resolutions of the candidate windows

The basic Haar features set is extended by Viola and Jones in [309] with two new feature types that emulate second order derivatives. The new set is known as *Haar-like features* set (Fig. 3.7). The addressed task is face detection. Additionally, this work introduces the use of the so-called *integral image* [56] for a fast and easy computation of Haar-like features at many scales. The key of integral images is that they allow to compute the sum of pixels of any rectangular region in just four memory accesses and three sums (Fig. 3.8). This implies that a given Haar-like feature can be computed in constant time at any scale and location. Accordingly, rather than using a pyramidal approach for generating candidates over different scales as in [232, 240, 241], Haar-like features are scaled themselves (Fig. 3.9). The speed up in the computation of the individual features allows Viola and Jones to start with a bag of around 180,000 features. Then, in order to reduce the number of features to compute during classifier operation, the training process focus on selecting a predefined number of the *best* ones. In particular, an AdaBoost methodology is followed, where the weak rules of the strong classifier (ensemble) are decision stumps (i.e., thresholding of individual features). In fact, such weak rules are structured as a cascade intended to reject easy non-face candidates at the initial layers, thus, accelerating the overall classification time. The complete face detection cascade has 38 layers with over 6,000 features and turns out to be a quite fast classifier. Viola and Jones' proposal has been included in the popular OpenCV library.[1]

This work is later extended by the same authors [160, 310] for detecting pedestrians in surveillance videos. A major contribution is not only using luminance appearance to characterize the pedestrians but also patterns of motion. Taking into account that in this application the camera is static, the background too, and the pedestrians are imaged from a relative zenital-like far distance, we can assume that from frame to frame they do not undergo large displacements in the image plane. Therefore, rather than relying on costly optical flow to account for motion information, spatio-temporal Haar-like features are used, based on two consecutive frames. Similarly, Ke et al. [163] use spatio-temporal basic Haar features for detecting human visual events. In [57] spatio-temporal Haar-like features are extended by Cui et al. to use more than two consecutive frames.

In [188], Lienhart and Maydt extend the Haar-like features with $\pm 45°$ rotated versions of the Haar-like features (except for the diagonal feature) and center surround features (Fig. 3.7). They provide also the algorithm for computing the new features with integral images. In [205], Messom and Barczak present Haar-like features rotated by almost any angle together with the method for computing their corresponding integral images.

Finally, it is worth to mention that different normalizations are usually applied to Haar-like features in order to gain robustness to different illumination, scale and background-foreground contrast sign (Fig. 3.7).

3.2.3 Edge Orientation Histograms Features

Having variability in the available set of features is desired in general. According to this idea, in [186], Levi and Weiss introduce the *EOH features* to complement Haar-like ones (Fig. 3.10). EOH features rely on gradient information, obtained according to a Sobel operator. The idea is to fix k different orientation bins and then, given a candidate rectangular sub-window, the gradient magnitude of each of its pixels is assigned to the orientation bin which is closest to the gradient orientation of the pixel. All gradient magnitudes assigned to a bin are summed. With the use of one integral image per orientation bin, it is easy and fast to compute the gradient magnitude summation of a bin for any candidate sub-window. Thinking in all bins we see that this procedure is an efficient way of computing weighted gradient orientation histograms for any candidate sub-window. Then, from this basic capability, different types of feature can be defined, for instance ratios between bins and dominant orientation (Fig. 3.10). Even symmetry measures can be defined by considering pair-wise regions placed symmetrically with respect to any desired axis of symmetry (e.g., the central column of the candidate window). As with Haar-like filters, varying the size and position of the candidate sub-windows, pools with hundred of thousand of EOH

---

[1] http://opencv.org

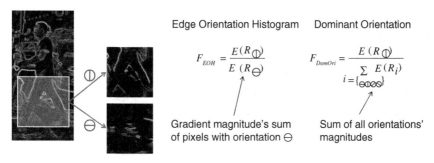

Fig. 3.10 Example of EOH feature computation withing a rectangular image region

features can be available for feature selection. Each EOH feature is a real value, thus, it can be easily used as decision stump to form a weak classifier for AdaBoost style algorithms. This is the approach in [186], where combining EOH features with Haar-like ones showed better accuracy than using Haar alone.

In order to build a pedestrian classifier, Gerónimo et al. [120, 122, 123] combine a sub-set of Haar-like and EOH features to form a feature pool mined by a Real AdaBoost algorithm with cascade structure. Similar work is presented by Chen and Chen [51, 52]. Differences between Gerónimo et al. and Chen and Chen works are the candidates generation process, the type of EOH and Haar features used to form the pool, and overall classifier structure. Candidates generation is 3D-based in the former works, 2D in the latter ones. The classification structure of Chen and Chen includes meta-stages at each layer of the Real AdaBoost cascade. A meta-stage can reject or accept a candidate window as normal stages do, but the meta-stages take the decision using the output of previous meta-stage and current stage, these two output values are feed to a linear SVM to take the classification decision.

As with Haar filters, basic EOH features admit different modifications. For instance, instead of assigning the gradient magnitude to a single orientation bin, it can be distributed to adjacent bins (those bounding the gradient orientation) according to bin interpolation; gradient orientation can be considered with contrast sign or not depending on the application, etc.

3.2.4 Histogram of Oriented Gradients Features

In [60], Dalal and Triggs introduce the *HOG features*, which nowadays are the most popular ones by far in the field of pedestrian and general object detection. Analogously to EOH features, HOG ones exploit gradient information too. Conceptually, HOG computation can be thought as having three major steps (Fig. 3.11). In [60] different possibilities were explored for each step, here we summarized the version that was finally chosen and later popularized.

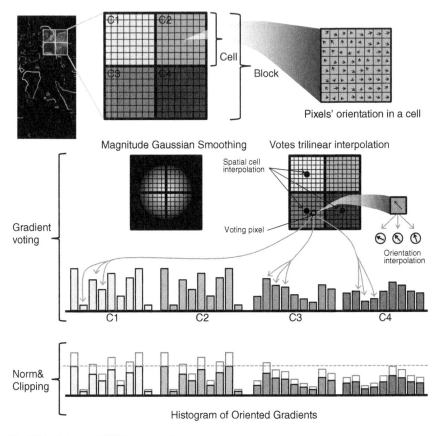

Fig. 3.11 Scheme of HOG computation

First, the gradient information must be obtained. Centered discrete differences are computed in [60] for R, G, and B color channels, then a per pixel *max* operation on the channels' gradient magnitude decides what channel gradient is used at each image pixel. Second, any candidate window is assumed to be composed of a grid of rectangular cells, 8×8 pixels (square) cells are used in [60]. Within each cell a histogram of gradient orientations is computed in a similar way to EOH, i.e., fixing k bins ($k = 9$ was chosen in [60]) and accumulating the gradient magnitude. In order to reduce cell-induced aliasing, interpolation is used for distributing gradient magnitudes to adjacent orientation bins and cells within the so-called block. A block is just a grid of cells, e.g., in [60] blocks of 2×2 cells are used. Within a block all the cell histograms are concatenated into what we could term as block histogram. These block histograms are then normalized, in particular [60] uses a L_2 normalization with hysteresis (i.e., analogous to SIFT descriptor [194] in which HOG is inspired). In fact, before computing the orientation histograms of the cells of a block, the gradient magnitudes of the pixels within the block are down-weighted according to a Gaussian filter centered at the block. By concatenating all block histograms of a candidate window we obtain the vector of features describing the window. In [60] the blocks within a candidate window overlap in one cell both in vertical and horizontal directions. Thus, each cell contributes four times, with different normalization, to the final window descriptor. Altogether, HOG features induce robustness against global and local illumination changes. Even moderate pedestrian pose differences (i.e., small relative shifts of body parts) are absorbed by HOG descriptor, as far as the local body shifts remain within the same underlying cells. In [60] a linear SVM is used to learn the pedestrian classifier based on HOG features. This is possible because the size of the canonical candidate window is of 64×128 pixels, thus, the final window descriptor has less than 4,000 features which can be handled by SVM.

Researchers also use HOG features to detect pedestrians in other modalities. For instance, Suard et al. [283] apply HOG to FIR images. Dalal et al. [61], analogously to Viola and Jones with Haar-like features, extend HOG to operate in videos. However, in [61] the camera is not assumed to be static, thus optical flow computation is needed. Several possible options of using HOG for optical flow are studied. In the ones with higher accuracy HOG features are separately computed for the vertical and horizontal components of the flow, and then combined with (static) luminance HOG features. Thus, HOG operates in two modalities, namely luminance and motion flow. Two different schemes are tested for doing the fusion, using a monolithic classifier and using an ensemble of classifiers. In the former, the static and motion HOG features are concatenated in a single vector and then a linear SVM is used to learn the pedestrian classifier. In the later, called *mixture of experts*, separated classifiers are learned from static and motion information, then the outputs of these two *experts* are used as features to learn a linear SVM based final pedestrian classifier. This type of ensemble is similar to the one used before by Mohan et al. [216] to build their parts-based pedestrian classifier (Sect. 3.3). Both fusion schemes provide similar results, but the mixture of experts is preferred due to its building modularity.

Aiming at accelerating candidate classification, Zhu et al. [342] propose a different scheme for exploiting HOG information. In this proposal HOG blocks can be of different size and aspect ratio, and at different location. For the used canonical window this turns out to produce around 50 times more blocks than original HOG proposal. HOG blocks are then used to learn linear SVM based weak classifiers from which an AdaBoost cascade is built. Additionally, block-level Gaussian weighting of the gradient magnitudes as well as interpolation to avoid cell-induced aliasing during gradient magnitude accumulation, are omitted with the argument of making possible the use of integral images, i.e., for using one integral image per bin as done with the EOH features. The step of histogram normalization relies on L_1 norm for speed acceleration too. Altogether lead to a pedestrian classifier exhibiting comparable accuracy to the one based on the original HOG, but being about 70 times faster. However, we note that in [315] it is shown that the interpolation operations involved in HOG computation are compatible with the use of integral images, and this is an important step to avoid losing accuracy. Later, Pang et al. [239] generalized such type of interpolation to sub-cell level. In [330], Xu et al. also use variable size HOG blocks for training a pedestrian classifier based on a cascaded L1-norm minimization procedure.

In [92] Felzenszwalb et al. carry out a PCA study of HOG features as defined by Dalal et al. As a result a set of new features of lower dimension are defined. These features demonstrate equal accuracy than original HOG and have the advantage of being computed in a more efficient manner.

3.2.5 Shapelet Features

EOH and HOG features are not the only ones trying to capture contour information. For instance, Sabzmeydani and Mori [266] propose the *shapelet features*. A shapelet is computed within a sub-window of the candidate window. A shapelet is defined as a weighted combination of contour features of the sub-window associated to the shapelet. Each contour feature has associated a location, a direction and a strength. In particular, a set of directions is fixed a priori (e.g., $0°$, $45°$, $90°$, $135°$ are considered in [266]). Then, the input image is separately derived in such directions and the results convolved with a box filter (i.e., per-pixel averaging in a square neighborhood). The resulting (four) averaged derivative images are used as source of contour features: the specific image derivative identifies contour direction, the derivative values contour strength, and the coordinate within the image sets contour location. Feature normalization is also used at shapelet level, in particular the derivative values are normalized within each shapelet sub-window considering all directions simultaneously. Per-direction integral images are used for fast normalization.

In order to obtain the weighted combination of contour features that form a shapelet, each contour feature is used as input for a decision-stump-based weak classifier that distinguishes pedestrians from background, and AdaBoost is used for selecting the weighted combination of such classifiers. Therefore, a shapelet is an AdaBoost-based strong classifier. Now, shapelets are used in turn as weak classifiers to learn a final strong pedestrian classifier using AdaBoost again. Since the shapelet is associated to a specific sub-window of the canonical window, this second round of AdaBoost learning has the purpose of selecting and weighting the relevance of such sub-windows. Seven sets of sub-windows sizes were tested: Small (5×5 pixels), Medium (10×10), Large (15×15), and combinations (SM, SL, ML, SML).

3.2.6 Local Binary Pattern Features

In [299], Ojala et al. introduce the original version of the *LBP features* aiming at locally describing 2D image textures. Such a version operates as follows (Fig. 3.12). For computing the LBP value at a given pixel, its raw value (e.g., luminance value) is compared with respect to the raw values of its eight adjacent neighbors, following a fixed visiting order. If a neighbor raw value is greater than the one of the central

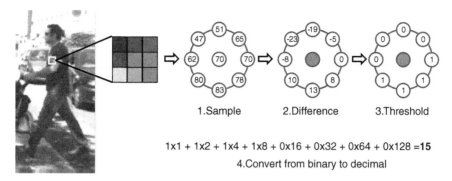

1.Sample   2.Difference   3.Threshold

1x1 + 1x2 + 1x4 + 1x8 + 0x16 + 0x32 + 0x64 + 0x128 =**15**

4.Convert from binary to decimal

Fig. 3.12   Computation of the LBP in a given pixel. Inspired by [251]

pixel, then it is marked as "0", otherwise as "1". The sequence of "0"/"1" values form an eight bit code easily convertible into a decimal one.

The basic LBP descriptor is generalized in [230], in particular the idea of *uniform* LBP is introduced. This is an important concept because it allows to map the values of a non-uniform LPB into a significatively reduced number of equivalence classes, which seem to be the most common in natural images. Let us call t the number of bitwise transitions from "0" to "1" or vice versa when the bit pattern is considered circular. A LBP is termed uniform if $t \leq 2$. For instance, the eight bits LBP values "11111111" (0 transitions) and "10000111" (2 transitions) are uniform, whereas the pattern "10101010" clearly is not. Accordingly, the original LBP with 256 possible values is mapped to 59 equivalent ones (including the class of non-uniform LBP).

Since the definition of LBP, this type of feature has been extensively studied and used in different applications. LBP for color and volumetric modalities has been also defined. For a comprehensive review the reader is addressed to a recent book by Pietikäinen et al. [251]. Indeed, one of those applications is pedestrian detection. In particular, Mu et al. [219] propose two modifications of the original LBP, the most successful called semantic LBP, which is used to operate in CIE-LAB color space. To describe a sub-window of a pedestrian candidate window, a histogram of the semantic LBP values is computed. In order to characterize the whole candidate window, a pool of sub-windows described with semantic LBP histograms is used. Such a pool is built by varying the size, location and aspect ratio of the sub-windows, i.e., analogously to what Zhu et al. [342] do with HOG blocks. Using the sub-windows LBP information a pedestrian classifier based on a LogitBoost cascade of rejectors is built.

After Mu et al., Wang et al. [315] combined LBP and HOG features. In this case, uniform LBP values are used and corresponding LBP histograms are computed according to the cell grid defined for HOG features. A LBP histogram obtained in this way is termed as cell-structured LBP. The HOG features are concatenated to the cell-structured LBP features in a vector feed to linear SVM algorithm in order to learn the pedestrian classifier. Moreover, as we have mentioned before, Wang et al. show how to adapt to the integral image mechanism the original HOG interpolation operations used to avoid cell-induced aliasing. In fact, as we will see later (Sect. 3.3),

Wang et al. propose an occlusion handling scheme too. Overall, the most important message is that the combination of HOG and LBP in feature space clearly outperformed the use of both types of features in isolation. This fact has been confirmed for general object detection as well, achieving state-of-the-art results [337]. Intuitively, HOG is capturing shape information whereas LBP contributes with texture information.

Other LBP variants have been also applied to pedestrian detection. For instance, for FIR images Sun et al. [285] modify the center-symmetric LBP to favor pedestrian-like symmetries (w.r.t. vertical axis), and a pyramidal approach is followed to account for symmetry at multiple levels. In fact, the descriptor is extended to capture motion information too. The spatio-temporal information is obtained following Viola et al. [310] style rather than a Dalal et al. one [61], i.e., the 2D descriptor is extended to 3D instead of relying in optical flow computation.

3.2.7 Dominant Orientation Template Features

In [138], Hinterstoisser et al. introduce the idea of *DOT features* for detecting textureless objects in real-time. DOT mainly focus on capturing gradient orientation information and codifying it in a very efficient way, i.e., using binary codes that can be obtained even from a single object instance. Thus, DOT shares ideas with templates, HOG and LBP. In fact, DOT is presented as an alternative to HOG in the sense of providing similar detection performance, probably (slightly) lower, but being even two orders of magnitude faster. In particular, the model of an object of interest is built as follows. The canonical window of the object can be divided into regions, i.e., like the cells of HOG. The dominant orientations of a region are defined as the gradient orientations associated to the k strongest gradient magnitudes over a threshold. All possible dominant orientations are discretized into seven bins, each bin has an associated bit within a byte code. Each region is represented by such a byte code. If any of the dominant orientations of the region belongs to bin b, then its associated bit is set to "1", otherwise to "0". The eighth bit is set to "1" if the region is considered uniform, i.e., if no dominant orientations were found, otherwise is set to "0". Jittered versions of the object example are considered during dominant orientation computation to obtain invariance to small 2D translations. By concatenating the bytes representing the object regions, a multi-byte representation of the object is obtained.

To decide if a given canonical window of an input image contains the object of interest, the DOT region splitting is applied to the window. However, only one dominant orientation per region is considered. In particular, the gradient orientation corresponding to the largest gradient magnitude of each region. The byte codes representing the regions are built and concatenated as above described. Then, comparing the object DOT with the image window under consideration is done through a bitwise AND. If 16 regions are used then SSE technology allows for very quick comparisons [138], i.e., the classifier can label the window as pedestrian or non-pedestrian very fast.

Robustness against scale, rotation, and viewpoint is obtained by using corresponding object examples to obtain different DOTs. Similarity between different DOTs representing different objects under different viewpoints is used to form cluster templates that allow for efficient branch and bound search [138].

Up to here, as the name suggests, DOT is more on the realm of the template features seen at the beginning of this section. However, as to the best of our knowledge, the first application of DOT idea to pedestrian detection is done by Tang et al. [288] following a discriminative learning scheme based on a random forest cascade. At non-leaf nodes of the forest's trees the split function is based on the distance between a pedestrian-patch DOT (no regions are used to further divide the patches) and an candidate-patch DOT. If the distance is higher than a threshold then the candidate patch goes down to, say, the right of the tree, otherwise to the left. In the process of training such nodes both pedestrian and background patches are used to learn the discriminative thresholds. In fact, DOT is not the only feature used in [288]. Since DOT has less discriminant power than HOG, more fast-to-compute features are combined with DOT. In particular, color images are assumed and HSV color space computed. Then, pedestrian patches are represented by a DOT byte and a color byte, i.e., by dominant orientations and dominant color.

3.2.8 Co-Occurrence Features

Summarizing the values of a feature in a given region (e.g., candidate sub-windows) using a histogram of its values is the core of descriptors like HOG. However, this methodology throws away spatial relationships among the feature values (e.g., symmetry is not explicitly modeled). In order to avoid this, *co-occurrence of feature values* can be considered. In fact, their use can be traced back to late 70s when such co-occurrences and statistics computed from them were proposed to characterize textures [65, 344]. The basic idea of co-occurrence of feature values is as follows. First, the feature values are discreticed (i.e., like to compute a histogram). Second, a desired spatial constraint is defined for any two discreticed feature values, for instance, *the two feature values must be at a distance lower than D pixels*. With these basic ingredients we compute the co-occurrence matrix for a given image region as follows. The index of the matrix columns runs on the discreticed feature values, and the same for the rows. Then, at any matrix entry we count the number of times that the feature values corresponding to its column and row indices accomplish the defined spatial constraint within the region under consideration.

This idea is applied to pedestrian detection by Watanabe et al. in [316]. In this work the feature is gradient orientation which in quantized in bins, i.e., as for HOG computation. Co-occurrence is also defined within regions that are equivalent to the cells of the HOG descriptor. The spatial constraint is given by a shift vector between feature values. In fact, several shift vectors are considered, thus, several co-occurrence features are computed per cell (Fig. 3.13). Accordingly, these features are termed as co-occurrence histograms of oriented gradients (CoHOG). All co-occurrence matrices are unfolded and concatenated into a very high dimensional vector that is used to learn a pedestrian classifier with linear SVM.

A major problem of this approach is the huge dimensionality of the feature vector, easily hundred of thousands. This turns in a high computational time since SVM requires all the features to be computed, and note that just a decade ago this dimensionality could not be even handled by SVM training. In order to cope with these problems, Hiromoto et al. [139] divide the CoHOG feature vector into many smaller ones and construct a rejection-cascade with weak classifiers based on such smaller vectors. However, CoHOG still throws away potentially relevant information, the gradient magnitude. Accordingly, Pang et al. [238] modify CoHOG to account for gradient magnitude during the co-occurrence voting process. For each matrix co-occurrence entry two gradient orientation are involved, so two corresponding gradient magnitudes too. Therefore, several strategies are considered to define the *amount of gradient magnitude* to accumulate. In particular, adding gradient magnitudes of the co-occurred gradient vectors, multiplying such magnitudes, as well as taking the magnitude of the vector resulting from the addition of the two co-occurred gradient vectors. The last approach performed best. Moreover, in order to favor high speed classifier execution, Pang et al. make use of integral images for anti-aliasing interpolations and perform a drastic dimensionality reduction based on incremental principal component analysis (IPCA).

Co-occurrence of features are not always measured using accumulation matrices. Another possibility is as follows. Each single feature is replaced by a counterpart binary variable that takes the value "1" or "0" depending on the feature value (e.g., thresholding the feature value or its absolute value). Then, a co-occurrence feature consists in combining several of those binary variables. For instance, if we combine three single features, the resulting co-occurrence feature is represented by a binary number of three bits. Then, given a labelled training set, we can count the instances of each binary number for the positive samples and, separately, for the negative ones, thus forming the corresponding likelihood functions (Fig. 3.14). Note that the idea behind LBPs is similar to the one exposed here, in fact, some authors define LBPs as co-occurrences of neighboring pixel pairs. Thus, the histograms of cell-structured LBPs [315] being cell-likelihood functions of the co-occurrence features (LBPs).

Mita et al. [210] follow this method of capturing co-occurrences for face and hand detection. The single features are Haar-like ones, from them a vast amount of co-occurrence features can be generated. Note that given an amount of single features to combine, many possible combinations are generated since the Haar-like features vary in position and scale. In addition, different amounts of combined features are considered. Each co-occurrence feature is used for computing a weak classifier within an AdaBoost learning framework (both Discrete and Real versions are tested). The weak classifier takes into account the likelihood functions for taking the classification decision. The selection of the best weak classifiers during AdaBoost learning is based on a greedy search in the feature space (sequential forward selection).

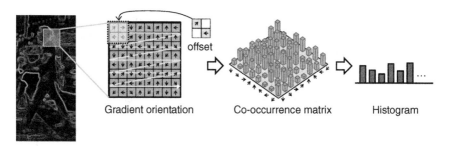

Gradient orientation   Co-occurrence matrix   Histogram

Fig. 3.13 Co-occurrence of features in the case of HOG for a given spatial realtionship (particular offset). Unfolding the matrix we obtain a one dimensional histogram. By defining other spatial relationships (e.g., other offsets) we obtain more co-occurrence matrices. Inspired by [316]

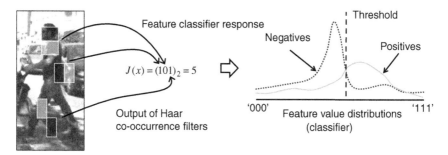

Fig. 3.14 Co-occurrence of features codified as binary numbers and corresponding likelihoods for the pedestrian and background classes. Inspired by [210]

Mitsui et al. [211] follow a similar approach for pedestrian detection. However, they use HOG as source of single features. More specifically a co-occurrence feature is defined by choosing several HOG cells (after block normalization) and one orientation bin per cell, the thresholding of each bin value (in absolute value or not) defines the bit values of the binary code of the co-occurrence feature. A two stage Real AdaBoost is used to learn the pedestrian classifier, where the first stage generates a pool of discriminative co-occurrence features and the second stage mines it to produce the final strong classifier. In [333], Yamauchi et al. propose the co-occurrence of probability features (CPFs) by using addition and multiplication of weak classifiers based on single features (HOG bin values as in [211]). A nested Real AdaBoost algorithm is also used for training a pedestrian classifier. An interesting novelty is the use of CPFs between pedestrian features and the geometric context (GC) features proposed by Hoiem et al. [141]. In particular, GC analysis assigns to each image pixel a probability value of belonging to *ground*, *vertical-object* and *sky* classes. These values are averaged per-class at the HOG cell level. Then, given a GC class, a new co-occurrence feature is computed by subtraction between a weak classifier based on a pedestrian feature and the average probability of the HOG cell corresponding to the pedestrian feature.

The co-occurrence idea is also underlying the proposal of Duan et al. [76], who propose the so-called *associated pairing comparison features* (APCF). In particular, the concept of *granule* is defined as any sub-window of the canonical candidate window, i.e., at any position and of any size. Comparison of color information between two granules lead to pairing comparison of color (PCC) features and, analogously, PCG features are obtained by using gradient information. Then an APCF is defined as a sequence of PCC and PCG features (can be mixed). The number of possible APCFs is enormous, thus, a feature selection algorithm is used to mine them.

In the same spirit, Cao et al. [47] use pair-wise relationships between HOG feature vector components, which will be later termed as *feature interaction descriptors* (FINDs) [126]. More specifically, HOG blocks are considered according to the classical 4×4 arrangement of cells. Then, given one of such blocks we can think of it as the vector resulting from the concatenation of the histograms of gradient orientations of its four cells (according to HOG parameters of binning, etc.), but before block normalization. Then, pair-wise relationships between the components of such a block vector are considered, i.e., following an all-versus-all approach. Such relationships are defined according to single-valued functions, then a block vector of n features is transformed in a $n \times n$ new vector of features (the single-valued functions no necessarily must be symmetric with respect to the two inputs). In [47], harmonic mean, min and product are tested as pair-wise relationships. Using linear SVM to build the FIND based pedestrian classifier, harmonic mean and min performed better than product.

3.2.9 Covariance Features

In [298], Tuzel et al. introduce *covariance features* as fast region descriptors for performing object detection and texture classification tasks. Later, the same authors use covariance features for pedestrian detection [299, 300]. The basic idea of covariance features is as follows. First, n *pixel-wise* feature types are chosen. For instance, in [299, 300] they consists of the pixel coordinates, the absolute value of its first and second order partial derivatives (except the cross one), and the gradient modulus and orientation (i.e., $n = 8$). Therefore, from a set of pixels within a given image region, an $n \times n$ covariance matrix of features can be computed. The diagonal elements of the matrix are the feature variances whereas the off-diagonal entries are correlations of features. Note that covariance matrices are symmetric, thus, they can be seen as a $(n^2 + n)/2$ dimensional feature vectors (36D for $n = 8$) representing the corresponding image region. Now, given a pedestrian candidate window, we can think of many rectangular sub-windows contained in it, each one represented by its corresponding covariance matrix. Moreover, the integral image technique can be applied to compute the covariance matrix of any image sub-window in a fast manner [298]. In [299, 300] it is defined an initial pool of overlapped sub-windows by considering a range of sizes for all pixels of the candidate window. Therefore, from a candidate window we obtain thousands of covariance matrices. In order to learnt a pedestrian classifier based on such a bunch of matrices a boosting with cascade structure is used. Since covariance matrices do not form a vector space but a Riemannian manifold, weak classifiers must operate in the tangent space of the manifold. In fact, rather than classifiers, weak regressors selected with LogitBoost algorithm are used. Each regressor is obtained by least squares regression and there is one per sub-window (i.e., per covariance matrix).

3.2.10 Data-Driven Features

So far, we have seen different manually-designed descriptors used as features by some machine learning algorithm to obtain a pedestrian classifier. Another approach consists in using the raw data or very low-level data cues (e.g., gradient magnitude) and learn the descriptors themselves as part of the process of learning the pedestrian classifier, i.e., learning data-driven features. This is the realm of artificial *neural networks* (NNs).

Zhao and Thorpe [341] use luminance gradient magnitude as input for a three-layer fully-connected feed-forward NN (*multi-layer perceptron*, MLP in short). Each neuron uses a sigmoid activation function. There are 30×65 neurons in the input layer (i.e., the size of the canonical window), five in the hidden layer, and one in the output layer. This last one is thresholded to decide if a candidate window contains a pedestrian or not. Stereo information helped to remove out-of-range information from candidate windows previous to their classification.

Rather than using a fully connected NN, Munder, Enzweiler and Gavrila [83, 220] evaluate the pedestrian classification accuracy of a three-layer feed-forward NN with *local receptive fields* (LRFs) in the hidden layer. Inspired by Hubel and Wiesel [148–151], LRFs were first introduced in [106, 107] for (visual) pattern recognition and used for gait-recognition-based pedestrian detection by Wöhler et al. [319, 320].

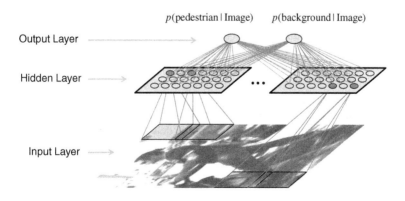

Fig. 3.15 LRFs/NN inspired by [83]

The idea of the LRFs/NN is that each neuron of the hidden layer is only connected to a local region of the (luminance) input image, its LRF. Additionally, the hidden layer is divided into branches (16 are used) and all neurons within the same branch share the same set of weights. Local connectivity and weight sharing reduce the number of network's parameters to learn, which a priori allows for smaller training sets than in the fully connected case. The output layer is composed by two neurons, their outputs correspond to the posterior probability estimates for pedestrian and non-pedestrian classes (Fig. 3.15). The experiments show that for a canonical window of 18×36 pixels, optimal LRF size is 5×5 with 2 pixels shift among adjacent LRFs, i.e., LRFs overlapping is necessary. Interestingly, Munder and Gavrila show that using the outputs of the hidden layer (i.e., LRFs outputs) as features for a quadratic SVM, the obtained classifier outperforms the result of using the whole NN. Radial basis function (RBF) kernel was also tested with similar results. Of course, the LRFs/NN training is performed anyway to be able to obtain the hidden layer outputs. However, the LRFs/SVM solution was discarded for excessive memory requirements during training [83]. In order to take into account multi-resolution pedestrian detection, Enzweiler and Gavrila [83] consider two canonical windows, the mentioned of 18×36 pixels (2-pixels frame, 5×5 LRFs, 2 pixels shift) and another of 40×80 pixels (4-pixels frame, 10×10 LRFs, 5 pixels shift).

Following the Hubel-Wiesel's multi-stage architecture inspiration, half way between classical MLP and LRFs/NN we find the *convolutional neural networks* (CNNs) introduced by LeCun et al. [177–179]. A CNN is a special case of MLP whereas a LRFs/NN is a special case of CNN. Basically, CNNs are more specific than MLPs in the same sense than LRFs/NNs, i.e., because the use of locally limited connections and weights sharing; whereas CNNs are more general than LRFs/NNs because they do not rely on three layers but usually they are composed of one to three stages, each stage composed of several layers. In particular, the last stage is usually fully-connected and acts as a classifier based on features (like in MLP), while the former stages act as feature extractors and can have three layers per stage, namely a convolutional filter bank layer, a non-linear transform layer, and a down-sampling layer.

Szarvas et al. [287] use a three-stage CNN for pedestrian detection. The canonical window is of 30×60 pixels (10-pixels frame). Interestingly, Szarvas et al. also show that the accuracy of the CNN is worse than using the features learned by the CNN to build a final classifier based on SVM with Gaussian Kernel. Szarvas et al. argue that the reason is that SVM directly performs a large-margin-keeping optimization and, therefore, a large margin CNN is proposed. CNN-features/SVM and Large-Margin CNN have a similar detection performance but the latter is about 40 times faster. We can find more recent examples of pedestrian detection based on CNNs showing state-of-the-art accuracy [162, 274].

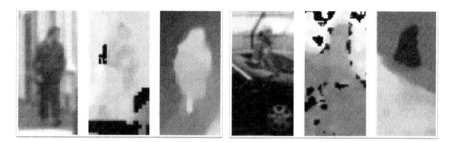

Fig. 3.16 Two pedestrians seen under different modalities (monochrome appearance, depth from stereo, and optical flow). Image courtesy of Dariu Gavrila (Daimler AG)

3.3 Diversified Models: From Features Fusion to Multiple Parts

In Sect. 3.2 we have seen how the different proposals to build holistic pedestrian models mainly focus on looking for discriminative features. However, in the pedestrian detection field this has not been the only research trend. For instance, there are approaches in which existing features are combined rather than designing new ones. In other approaches, pedestrian classifiers of different nature are ensembled or cascaded. While in many holistic approaches pedestrian examples showing different pose and view are just mixed to learn the pedestrian classifier, some works do multiple views training, multiple (body) parts training, or both. Special difficulties such as having pedestrians at multiple resolutions or being partially occluded have been analyzed too. In this section we summarize the main works covering these research lines.

3.3.1 Combined Features

Regarding the fusion of different feature types we can find two main strategies. On the one hand, given the same type of information modality different image cues can be collected by different descriptors, i.e., the diversity comes from the descriptor. On the other hand, we can apply the same type of descriptor to different modalities, i.e., the diversity comes from the modality (e.g., Fig. 3.16). Of course, both things are also possible.

In Sect. 3.2 we have already seen examples of combining different descriptors using a single modality. For instance, we have mentioned the concatenation of HOG and LBP features that are feed into linear SVM [305]. Usually it is said that HOG captures shape information while LBP captures texture information. Similarly, we have mentioned the combination of EOH and Haar-like features in a pool mined by some AdaBoost variant [51, 52, 120, 122, 123, 186]. In [322], Wojek and Schiele evaluate several combinations of dense shape context (descriptor used in the ISM template model), HOG, and Haar-like features using both linear SVM and AdaBoost with decision tree stumps as weak classifiers.

In [72], Dollár et al. propose the *integral channel features*, which combined with a boosting and soft-cascade algorithm [336] provide state-of-the-art results. A channel of an input image is another image of the same spatial dimensions and single-valued pixels. Such values are obtained from any local/global linear/non-linear transformation of the input image. For instance, a RGB color image is formed by three color channels, the gradient magnitude of a luminance image also provides a channel, the binary edges of an image provide another channel too, and so on. By using the integral image strategy for fast computation of regional sums, channel features of different order are defined. A first order channel feature corresponds to the summation of values over a rectangular region of a single channel. A higher order channel feature corresponds to any feature that can be computed using multiple first order features (e.g., the difference of two first order features of the same channel is a second order one). Note that the idea is similar to Haar-like and EOH features. In [72] there are tested two types of channels, namely, gradient channels (magnitude and bins of the orientation histogram) and color channels (luminance, R-G-B, H-S-V, L-U-V). From these channels, features are generated randomly. More specifically, channels are selected at random, first order features are generated by randomly selected rectangles (with a minimum area of 25×25 pixels), and higher order features are randomly generated weighted sums of first order ones (note that mixing channels is allowed). In order to mine the very high number of generated features, boosting is used (AdaBoost, Real AdaBoost and LogitBoost performed similarly) with trees of depth two as weak classifiers. Overall, first order features of gradient magnitude plus L-U-V channels provide the best accuracy.

Channel features are also an example among different proposals that try to incorporate color information into pedestrian detection. In Sect. 3.2 we also mentioned the combination of DOT features with HSV color information [288]. Ott and Everingham [234] combine HOG and so-called CHOG to obtain an enhanced feature vector feed to SVM (linear and quadratic kernel tested). In order to compute CHOG, first the RGB input image is pixel-wise normalized by the maximum color channel value at each pixel. Then, a fast soft-segmentation transform is performed at each block, which basically enhances background-pedestrian gradients. Finally, CHOG consists in apply HOG to the result of such soft-segmentation.

In [270], Schwartz et al. combine local texture statistics (following [135], 12 statistics are used) of co-occurrence matrices computed from HSV color channels, HOG features using the type of blocks considered in [342] (Sect. 3.2), and color frequency per-block. As we mentioned in Sect. 3.2 regarding HOG computation, at each pixel it is considered the gradient information corresponding to the color band with the highest gradient magnitude. Then, color frequency is defined as the number of times each color band is selected. Since the joint number of features is very large (e.g., almost 180,000 features for a 64×128 canonical window), a classifier based on partial least squares (PLS) analysis is proposed, which drastically reduces the dimensionality (20 dimensions) without losing discriminative information.

In [155], Ito and Kubota extend the CoHOG idea (Sect. 3.2) to work with color information. In particular, three sets of features based on co-occurrences are defined, namely, color-CoHOG (co-occurrence of the same color, and co-occurrence of different colors), CoHED (co-occurrence histograms of pairs of edge orientations and color differences) and CoHD (co-occurrence of pair of color differences). CoHED takes into account the relationship between and edge orientation and the change of color across the edge. CoHD accounts for changes of color values of three pixels located in a given line within the image. Color histograms and second order terms based on them are also used to complement the co-occurrence features. YCbCr color space is used as well as a linear SVM implementation able to cope with high dimensional spaces.

Anwer et al. explore the use of a biologically inspired color space for pedestrian detection using both a holistic [9] and a parts-based [8] pedestrian model. More specifically, the opponent color space is used. On top of it, HOG is applied to the three bands of this space (one accounts for luminance and two for chrominance) and the HOG features from each band are concatenated to form the final features vector. For the holistic model linear SVM is used to learn the classifier, whereas for the parts-based model the choice is Latent SVM [92, 93]. Anwer et al. focus on the relevance of color depending on the scenario. It is shown that in landscape scenarios the detection performance using color improves over using only luminance, in city scenarios it does not seem to be any improvement, while in ones the improvement is very narrow.

In [314], Wang et al. combine HOG with so-called max dissimilarity of different templates (MDST) computed on a color space (CIE-LUV and CIE-Lab are recommended according to the obtained results).

Walk et al. [312] also study the combination of color based features not only with HOG but with histograms of flow (HOF). The color features are based on *color self-similarity* (CSS) between different blocks of the canonical window, i.e., they rely on second order statistics (co-occurrence style) of color. The rational behind them is that while specific colors vary a lot within the pedestrian class as well as across the background one, color exhibits many times local repetitiveness. For instance, although defining an universal color for pedestrians' T-shirts is not possible, we know that a great part of a particular T-shirt may exhibit the same color (whatever it is). Different color spaces were tested (RGB, HSV, HLS, CIE-LUV, normalized rg, HS and uv), and HSV selected as the best. To compute color self-similarity between any two blocks, a $3 \times 3 \times 3$ histogram on HSV space is computed per cell and then histogram intersection is used as similarity measure between two of such histograms, and so between the corresponding blocks' colors. All blocks are compared and the self-similarity values appended into a feature vector that is normalized before using it for learning. In Sect. 3.2 (see HOG) we already mentioned that Dalal et al. combine HOG features computed on top of the optical flow with standard luminance HOG features [61]. Walk et al. combine CSS, HOG and HOF. Regarding HOF, Walk et al. use a lower dimensional variant of the Dalal et al. proposal. For learning the pedestrian classifier, histogram intersection kernel (HIK) SVM and MPLBoost are tested, the former being finally preferred.

Dalal et al. and Walk et al. proposals are not the only combining features coming from motion flow and luminance information. In Sect. 3.2 we already mentioned the use of Haar-like descriptors on top of both static and spatio-temporal images by Viola et al. [310]. Wojek et al. [325] follow an approach similar to Dalal et al. regarding the combination of luminance and optical flow HOG-style features, but they also consider luminance Haar-like features for the combination. Linear-SVM, HIK-SVM, AdaBoost, and MPLBoost are tested as learning machines. Haar features do not provide a significant improvement in the performed evaluation, and HIK-SVM and MPLBoost show similar detection performance, being better than the one provided by Linear-SVM and AdaBoost. An interesting observation is that motion based features offered specially good information for pedestrians imaged in side view, more than for frontal/rear views, which is totally in agreement with the intuition in the sense that the perceived motion for side viewed pedestrians is more clear (horizontal motion flow in image coordinates) than for rear/frontal viewed ones (flow goes into or away from the image plane, i.e., small optical flow vectors).

3.3.2 Classifier Cascades/Ensembles

Instead of combining different features, or in addition to that, a way of having more diversified pedestrian classifiers consists in cascading or ensembling them. In fact, we have reviewed examples of such approach since AdaBoost variants rely on weak pedestrians classifiers, over homogeneous or heterogeneous feature sets, to learn a strong pedestrian classifier. Note, that the adjective *weak* is used for classifiers whose accuracy is slightly better than random, it does not refer to the inherent complexity of the method used for learning the classifiers. For instance, weak classifiers can go from a simple decision stump until some SVM variant. In addition AdaBoost variants can produce rejection cascades (pipelined classifiers) as well as (linear) ensembles. Here, we focus on pedestrian detection, for an extensive review on combining pattern classifiers the reader is referred to L. I. Kuncheva work [172].

In order to improve the detection performance of the chamfer system (Sect. 3.2, Templates), Gavrila [111] adds a second classification stage. In particular, detections coming from the chamfer system are taken as candidate windows, thus, they are normalized to a canonical size and processed by a classifier based on a radial basis functions (RBFs) NN. For detecting pedestrians in FIR modality, Mahlisch et al. [196] use a chamfer matching and an AdaBoost cascade based on Haar-like features (a hypermutation network together with a relaxed flat word assumption generate candidate windows in this case).

Paisitkriangkrai et al. [237] use a two-stages cascade with heterogeneous features. In particular, the first stage is based on fast-to-compute features to favor speed, whereas the second stage is based on more accurate but slower-to-compute ones. More specifically, Haar-features are used for the first stage and covariance features for the second one. AdaBoost is used to build both the first and second stage strong pedestrian classifiers.

In [231], Oliveira et al. propose an ensemble architecture in which the (normalized) outputs of three parallel strong pedestrian classifiers are combined by a Sugeno fuzzy integral that provides the final pedestrian/background decision score. One of the strong classifiers is based on HOG and SVM, a second one is based on LRFs (computed trough a previous CNN learning) and MLP, and the third one uses the same LRFs with SVM.

Xu et al. [331] build a tree classifier ensemble where each node is an AdaBoost strong classifier computed from Haar-like features. The parameters of the tree are explicitly optimized to favor speed without losing detection performance. In [332], Xu et al. focus on sudden pedestrian crossings and proposed three-stage pedestrian detection framework, the first two stages perform per-frame detection. The first stage (local) uses LBP differences to output image windows containing possible pedestrian motion. The second stage (frame) classifies such windows by using two cascaded half-pedestrian (w.r.t. the vertical axis) classifiers, the first based on HOG/Linear-SVM and the second on Haar/Linear-SVM.

In [132], Guo et al. also propose a two-stages pedestrian classifier. The first stage relies on Haar-like features and AdaBoost, while the second stage combines a heterogeneous set of features and RBF-SVM. The features of the second stage correspond to texture information (luminance energy, entropy, contrast, local stability), symmetry, edge information (invariant moments of Canny edges), as well as histogram of gradient orientations (three shape-driven sub-windows are defined, top plus bottom left and right, to obtain seldom histograms).

In [263], Rohrbach et al. fuse classifiers based on luminance and depth modalities, where HOG/Linear-SVM and LRFs/NN are tested as base classifiers as well as different fusion rules (sum, max, product, SVM).

3.3.3 Multiple Aspects

The combination of features and/or classifiers brings diversity to the overall pedestrian model with the aim of gaining discriminative power with respect to the background. Another way of increasing discriminative power consists in explicitly taking into account the intra-class variability of the pedestrians. A straightforward idea is to use a different holistic model for each main view and pose rather than using a single holistic model, which tends to be more blurred (Fig. 3.17). Such idea can be combined with the use of different features/classifiers. We term *aspect* to a specific configurations of view and pose.

We have already seen examples aligned with this idea because, in general, models based on templates (Sect. 3.2, Templates) must explicitly take into account the aspect: the pedestrian intra-class variability must be well captured within the set of templates. For instance, in the chamfer system [110, 112] the upper level of the hierarchy mainly account for differences in the view whereas the mid-low levels account more for pose differences within the views. In fact, in [114], Gavrila and Munder use a pedestrian classifier with a first stage based on the chamfer system and a second one based on LRFs/NN. Interestingly, the second stage relies on a set of pedestrian classifiers that are trained under the influence of the chamfer system. For all such classifiers it is used the same pedestrian and background training set. However, each classifier has an associated aspect cluster. The considered clusters correspond to the top level of the silhouette hierarchy used in the chamfer system, i.e., they are mainly view oriented. Then, during the training of a classifier, the influence of each training example is weighted by its membership to the cluster associated to the classifier, which is determined by submitting the example to the chamfer system. During detection, it is applied an analogous process. Each classifier is considered an *expert* (i.e., an aspect expert) and the overall system is seen as a *mixture-of-experts* approach.

This idea has been progressively augmented by: (1) taking into account not only shape (chamfer system) and texture (LRFs/NN) cues but also depth from stereo, as well as per-cue temporal transitions between body aspects [221]; (2) estimating the pedestrians' orientation in an integrated manner with their classification [84]; and (3) defining a probabilistic pedestrian classification framework based on 24 trained classifiers (experts) that account for four different views (front, left, back, right), three different modalities (luminance, depth based on stereo, horizontal component of the optical flow), and two sets of features (HOG and LBP) computed per-view on top of the different modalities treated as grey-level information (MLP with a single hidden layer is finally preferred for learning) [85].

The ISM (Sect. 3.2, Templates) has been also evolved to account for different aspects by Seemann et al. [272], it is the so-called *4D-ISM*. The aspect is the new dimension (4th one) of the ISM voting space. The most prominent aspects are learned by clustering according to pedestrian silhouettes. Agglomerative clustering with global chamfer distance as similarity measure is used. The silhouettes are extracted from video sequences taken with a static camera in which a motion-segmentation algorithm is applied, basically assuming one imaged pedestrian per frame. Interestingly, the different aspect models share codebook entries. In [273], Seemann et al. improve the 4D-ISM model by allowing the sharing of *local context*. In particular, not only global shape is used to assign the codebook entries to an aspect or another

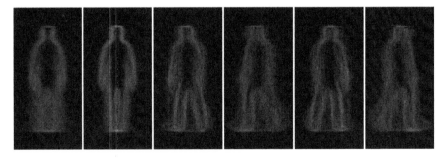

Fig. 3.17 *Left* overall silhouette average. The rest correspond to view specific silhouette averaging. Note how the overall model is more blurred

as in 4D-ISM, but the local appearance around interest points is also considered to perform accurate assignments across aspects. This local appearance, of controllable radii, is the mentioned local context. Both silhouette edges and a modified shape context descriptor are tested, being the former slightly better.

In [144], Hou et al. propose a rejection-cascade based on inspired HOG features and multiple view classification. The features capture dominant gradient orientations at local level, where *local* refers to either rectangular sub-windows or sub-windows with irregular shapes built by combining the rectangular ones. The sub-windows are called block in analogy with HOG, and 27 dominant orientations per block are considered. The rejection cascade itself has three stages. The first stage is a standard rejection-cascade segment where only rectangular blocks are considered in the firsts layers for speed reasons. The second stage classifies the pedestrians according to three views, namely frontal/rear, left and right. Pedestrian can be assigned to more than one view. The third stage has an specific rejection-cascade segment per view, thus, there are three of them. Hence, final classification decision is obtained according to such view-dependent rejection cascades. The learning methods used are Real-Adaboost and an extension of it, the Vector Boost, which is key for the second stage. In fact, this approach is based on a multiple-view approach to face detection [146].

Tran and Forsyth [295] also apply the idea of estimating the aspect and, conditioned to it, perform the final pedestrian classification. In this case the aspect is modeled according to a seven-points (six segments) wire skeleton encoding the position of the head (one segment), torso (one segment) and legs (two segments per leg); arms are excluded since they tend to be difficult to localize. Aspect estimation relies then on gradient-based features (PCA of vertical and horizontal gradients) extracted from rectangular patches defined according to the points of the wire skeleton. These features are used as a framework based on structure learning and dynamic programming to perform the aspect estimation. The aspect-estimation features and segment-segment layout features (segments' length and angle between them given the found aspect) are concatenated with HOG features (also of reduced dimensionality according to PCA) for training a SVM-based final pedestrian classifier.

Ye et al. [334] address the multiple aspect problem by manifold embedding and *error correcting output codes* (ECOCs) ensembling. More specifically, HOG features are computed and projected to a low dimensional manifold but preserving neighborhood relationships. Then K-means algorithm based on a geodesic distance clusterizes the pedestrian features within the manifold, the background features form an additional class (K+1th). Given a cluster, its base classifier is built according to linear SVM learning, not only using the examples of the cluster as positives but also the examples of neighboring clusters according to k-nearest-neighborhood rule. All negative samples are used as counterexamples of all learned base classifiers. Finally, the basic classifiers are ensembled by ECOC technique. Overall, the idea is that the manifold clusters correspond to different relevant aspects. Coarse-to-fine processing is also included for speeding up the classification of candidate windows.

Fig. 3.18 *Left* several holistic models to cover a set of aspects (e.g., see Fig. 3.2). *Right* a deformable parts-based model to cover the same aspects. The *boxes* frame the considered parts, which are arranged according to a plausible layout acounting for a range of shifts (star structure in this example) with respect to a reference (*anchor*) point (*circle*)

3.3.4 Multiple (Body) Parts

The use of several holistic models to cope with multiple pedestrian aspects has not been the only proposal in the literature. An alternative consists in the use of so-called (body) *parts-based models*, whose seed idea is an old one in the computer vision field [95, 98, 140, 153, 201]. In general, these models try to detect pedestrians' constituent *parts* and build a final pedestrian classifier ensemble, possibly combined with a holistic (full body) detector usually termed as the *root*. These are either direct body parts such as head-shoulders, trunk, arms, and legs, or at least body-inspired parts as when splitting the canonical window into up, mid, and bottom sub-windows. Parts-based modeling can be combined with previous approaches such as taking into account different major pedestrians views (i.e., different roots), using different features, etc. The body parts can either be assumed at fixed locations or searched in a range of allowed locations, this second approach, known as *deformable*, is supposed to be suited for detecting poses unseen during the training phase of the pedestrian model (Fig. 3.18) and remove spurious part detections. As an example, Fig. 3.19 shows a holistic model, a multiple views model, and a deformable multiple views parts-based model. Figure 3.20 shows detection examples based on such models.

Mohan et al. [216] propose parts-based detection as an evolution of the holistic model of Oren et al. [232] and Papageorgiou et al. [240, 241]. Haar features and a quadratic SVM are used to independently classify four human parts (head-shoulders, waist-legs, right and left trunk-arms). For obtaining the final pedestrian classifier, several ensembles are tested to combine the classification scores of the part detectors. Among them, linear SVM is finally chosen. Position constraints regarding the parts are considered during detection, both manually engineered and learned from the training data. Multiple views are not considered, although the approach can be extended to do it.

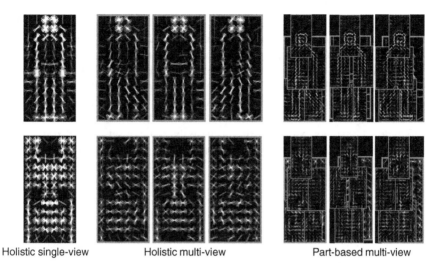

Holistic single-view   Holistic multi-view   Part-based multi-view

Fig. 3.19 Different pedestrian models trained by using HOG features and linear SVM. *Top* positively weithed gradient directions in the linear pedestrian classifier. *Bottom* negatively weithed gradient directions. *Left* single-view case. *Middle* multiple view case (left, frontal/rear, right). *Right* deformable multiple-view parts-based case (five parts used: head-shoulders, left arm-trunk, right arm-trunk, left leg, right leg) plus root model per view (occluded by the parts)

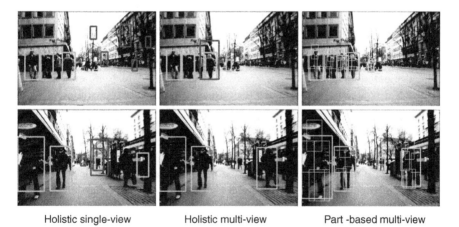

Holistic single-view   Holistic multi-view   Part -based multi-view

Fig. 3.20 Pedestrian detection examples according to the models in Fig. 3.19. *Dark boxes* are false alarms and *light boxes* are right detections. The detected parts of the parts-based case are also shown as *small light rectangles*

Shashua et al. [276] use thirteen sub-windows placed in a body-inspired manner (head, trunk, legs, head-trunk, trunk-legs, etc.), which are of different size and present overlapping. In fact, four of the sub-windows even come from concatenating four selected pairs of more *basic* sub-windows. Each sub-window is described by a sort of simplified HOG (note that [276] was presented before than HOG [60]). In particular, each of the nine basic sub-window is divided into 2×2 cells, each one described by a 8-bins histogram of gradient orientations whose counts are weighted according to a smooth version of the gradient magnitude. This results in a 32D vector descriptor per each of such sub-window. The descriptor for any of the four composed sub-windows comes from concatenating the descriptors of corresponding constituent sub-windows, thus, these are 64D descriptors. Then, ridge regression is used to learn sub-window classifiers. Moreover, in this case the training set is divided into nine clusters according to aspect and illumination conditions of the training pedestrians, resulting in $9 \times 13 = 117$ of such classifiers. The outputs of these classifiers are used afterwards as weak rules for building a final strong classifier according to AdaBoost algorithm.

In [293], Tosato et al. also present an approach in which parts are defined at different size levels (hierarchy of parts) and can overlap. The first level is the full body. The second level is a body-inspired partition of the candidate window into three parts, namely, head-shoulder, torso and legs. The third level takes into account seven semantic body parts, namely, the head, the left and right shoulders, arms and legs. In total, there are 11 parts taken into account. The parts classifiers are based on covariance features and LogitBoost (see Sect. 3.2). These classifiers are ensembled as a weighted summation rule learned from the training data. Multiple views and deformable parts layout are not considered.

Mikolajczyk et al. [209] propose a seven-parts-based (frontal head and face, lateral head and face, frontal upper body, profile upper body and frontal legs) human detector. The features describing each part are based on weighted dominant orientation, both computed according to gradient and Laplacian information, and co-occurrences of such orientations. Overall there are four feature types which are computed at five scales. Likelihood ratios of such features are used as weak classifiers for Real AdaBoost learning, i.e., to obtain strong classifiers. In fact, for each part, a strong classifier is computed per scale and feature type, thus, 20 of them are obtained. Such strong classifiers are further used to build a rejection-cascade per part. The final pedestrian classifier takes into account the results of the parts detectors as well as the relative position and scale between the parts, where each of such part-to-part geometric relationship is represented by a Gaussian whose parameters are learned from training data.

Along a series of papers [189–191], Lin, Davis et al. have proposed a hierarchical multi-aspect parts-based template matching for simultaneously detecting and segmenting pedestrians. A pedestrian silhouette hierarchy is learned, but rather than using full body aspects as in the chamfer system (Fig. 3.5), in order to cover the desired aspects, the tree is based on three parts-oriented layers given by head-torso templates (top layer), torso templates (intermediate layer), and legs templates (leafs layer). The connections between the nodes of adjacent layers are defined according to learned body aspect statistics. Thanks to this hierarchy and its associated template matching procedure, aspect-adaptive features can be used. First, HOG features are computed for a given candidate window. Then, these features are used to perform the template matching with the silhouette tree. Finally, HOG blocks are again computed for the closest regions of the matched silhouette, serving as features for a RBF-SVM based classifier.

It is worth to mention also the work by Wu and Nevatia [326–329] who study the performance of short segments (up to 12 pixels long) of lines or curves, referred to as *edgelets*, as features for multiple-aspect parts-based pedestrian classifiers. The application context, however, is more related to surveillance than to driver assistance, even when working with FIR images [340].

Fernández, Parra et al. [94, 245] propose a five-parts ensemble (head, left and right trunk-arms and legs, space between legs). Each part is modeled according to features based on co-occurrence of edges, histograms of gradient information, and texture unit number. These features are fed into a SVM to learn each part classifier. During detection, a majority vote approach is followed to take the final pedestrian versus non-pedestrian decision. In this approach, parts' location is assumed to be fixed and no multiple view considerations are taken.

In [223], Nanni and Lumini propose a body-inspired parts-based pedestrian detector. In particular, the candidate windows are divided into upper and lower parts, i.e., head-trunk-arms and legs. Previous to the part analysis, the candidate window is luminance normalized. Then, each part undergoes the same analysis, i.e., two types of features are computed and a RBF-SVM trained from each type. Thus, there are four classifiers to ensemble, two RBF-SVMs from the upper body part and two more from the lower body part. Following Kittler et al. [168] the *sum rule* is tested as ensemble method, outperforming other tested ensembles. One of the feature sets is obtained by computing an invariant LBP histogram (ten bins) for each output of the convolution of the part sub-window with Gabor filters. A bank of 16 filters is used by varying scale and orientation parameters. The other set of features corresponds to the Laplacian EigenMap (LEM) of the sub-window (a non-linear dimensionality reduction technique, which in this case is set to output 100 dimensions/features).

Yu et al. [335] propose a detector ensemble based on Dai et al. work [58], i.e., although multiple view considerations are not explicitly taken, we can say that co-occurrence of parts is considered as follows. The classification process is based on two main stages. In the first stage a holistic pedestrian classifier based on Haar-like features and AdaBoost is applied. Candidate windows classified as pedestrians in this first stage go to the second stage, which focuses on parts-based classification. The considered parts are head, torso, legs, left arm and right arm. The parts classifiers are built using shapelets and AdaBoost. Moreover, each part is searched around a small region where it is more likely to appear (the head around the candidate window upper area, and so on). Then, the final classification decision is taken according to an ensemble of the so-called substructure. Substructures are formed by three parts, and parts can be shared by different substructures. A substructure is said to be valid if all its parts are detected and a signed decision function that takes into account the relative locations and scales of the detected parts also provides a positive result. The final ensemble of substructures outputs a positive (final) classification response if some substructure is valid and no substructure obtains a negative response from the decision function. The combination of parts to form substructures as well as the number of substructures for the ensemble are optimized for the training data.

Huang and Nevatia [147] also extend to a multiple parts approach the work of Duan et al. [76] based on the APCF co-occurrence features. In this case, a single level of body-inspired parts is used. In particular, the canonical candidate window is divided into upper and lower sub-windows as well as left and right sub-windows. Thus, the parts are assumed at fixed positions and no multiple views are considered. The final pedestrian classification decision is taken in a Bayesian framework which incorporates the parts classifiers.

Felzenszwalb et al. [92, 93] have presented one of the most widespread parts-based object classifier up to date, which shows very good results on the pedestrian classification task too. The method does not assume a fixed position for the parts but a learned spatial layout of them, thus, it is a deformable parts-based model. On the other hand the number of parts and their size are fixed parameters. Moreover, the human body as a whole, termed as the *root*, is also considered (fixed size). Notably, the parts are modeled using twice the image resolution than the root. Root and parts features as well as spatial layout configuration determine the final classification score. The used features are HOG-inspired and a special SVM algorithm, called *latent SVM*, is applied. The model also allows to account for different views (one root per view), which are automatically determined according to the bounding box aspect ratio of the training pedestrians. Basically, side viewed pedestrians can be distinguished from frontal/rear viewed ones. The final location of the parts allows to compute the overall location of the detected pedestrians, which tends to be more precise than using the root detection.

Intuitively, latent SVM is solving a chicken-egg problem, namely, having parts classifiers we can locate the parts while located parts are required to learn the parts classifiers. Thus, if manual annotation of parts is to be avoided, the process involves an initialization step as well as an iterative optimization stage. During the initialization, each root classifier is trained separately using the HOG-inspired features and linear SVM. Then all initial root classifiers are combined into a mixture model with no parts. For each root, the position of its parts is heuristically initialized. Then, the parts classifiers are learned by using twice the image spatial resolution (i.e., the resolution at which HOG-inspired features are obtained) of the training pedestrians than the one used for the respective root. Thus, the initialization step provides a first current model (roots, parts, and spatial layout). Then, the second stage iterates the application of the current model to the training set with the updating of the model parameters (classifiers and parts deformation cost) as a consequence of the obtained results. The iteration runs until model stability or a maximum number of iterations is reached. A very important consequence of this process is that even though parts are not labeled a priori (i.e., using a bounding box), parts alignment is maximized at each iteration, which has been demonstrated to actually improve results [70] as intuitively expected. Figure 3.21 shows an example of a single-view model learned by this approach.

We would also like to mention the concept of *poselet* introduced by Bourdev et al. [35, 36]. The poselet-based method provides simultaneous person detection and segmentation. In fact, it is very time consuming and results presented in [35, 36] do not include ADAS oriented pedestrian detection datasets, but they do include

Fig. 3.21 Parts-based pedestrian model and detections [93]. From *left* to *right* HOG model of the root; HOG model of eight parts; spatial layout cost function of the parts (the darker the less deformation cost penalty); detections in Daimler A.G. dataset (see Chap. 5), shown boxes frame the detected root and parts

very diversified and complex poses. Poselet model is conceptually more evolved than the one proposed by Felzenszwalb et al. [92] in the sense of having greater flexibility and, thus, is expected to have superior accuracy when more training data is available. This last point is important since bounding box annotation of pedestrians is not sufficient, instead body key points must be provided. The body key points of a poselet determine the local aspect in our terminology, and the poselet appearance is modeled according to HOG features and linear SVM. Spatial layout statistics between poselets are also learned and taken into account. In a way, poselets can be though as arbitrary body parts under a particular aspect. Basically, for pedestrian detection, each poselet detector must be computed at every position and scale of the input image. Thus, poselets' spatial layout coherence forces plausible full-body aspects and avoid mixing parts coming from different persons.

It is also worth to mention the proposal of Andriluka et al. [7], which shows the ability of dealing with very challenging aspects similarly to poselets. In fact, pedestrian detection on ADAS-like images as well as pose estimation in, for instance, TV images, is addressed under the same probabilistic framework. The approach relies on a kinematic tree prior of the human body as well as part classifiers based on shape context features and AdaBoost with decision stumps.

3.3.5 Multiple Resolutions

Multiple views is not the only factor of variability introduced by the process of imaging pedestrians. Another important factor is the distance from the camera to the pedestrians, since in the ADAS context there are pedestrians observed at different distances within the same image (Fig. 3.4).

Fig. 3.22 Four multiple resolution approaches, figure based on [18]. From *left* to *right* (*direct* approach) a different classifier per resolution of detection (dense classifiers pyramid), image features are computed only at the original resolution; (*dense image pyramid* approach) a single resolution classifier and multiple resolutions of the input image (dense image pyramid), image features are computed for each resolution; (FPDW approach) a single resolution classifier and a sparse pyramid of the input image, image features at intermediate resolutions are interpolated from the ones at the explicitly considered resolutions (i.e., the sparse pyramid); (VeryFast approach) a few classifiers explicitly trained for seldom resolutions (sparse classifiers pyramid), classifiers at intermediate resolutions are derived, image features are computed only at the original resolution. The trade-off to address is the number of pedestrians classifiers to train (off-line) versus pedestrian detection speed due to multi-resolution feature computation in the input image

The immediate consequence of dealing with multiple distances is that pedestrians must be searched at multiple resolutions within the images. Multi-resolution detection has been traditionally addressed either by scaling the features (i.e., generating a pyramidal representation of the pedestrian model features from the one learned according to the canonical window, e.g., see Fig. 3.9 for the Haar-like ones) or scaling the images (i.e., generating a pyramidal representation of the whole image, see Fig. 3.22). In our experience with Haar-like and EOH features, scaling the images provides much better accuracy although the overall detector is slower. More specifically, we use to apply features scaling [120, 122, 123, 254] following Viola and Jones [309], but since more recently we use image scaling [306] with such features and we have obtained accuracy improvements of even more than 30 points in the per-image evaluation (Sect. 5.2) of some well-known data sets (e.g., INRIA and Daimler-Detection, Sect. 5.1). Notably, Dollár et al. [71] developed a hybrid approach for their integral channel features (i.e., computed using integral images), termed as *fastest pedestrian detector in the west* (FPDW), in which a sparsely sampled image pyramid is constructed with a resolution step of one octave, and channel features pyramids are constructed within each octave by following an exponential scaling law based on natural image statistics (Fig. 3.22). Accuracy is maintained with respect to a dense image pyramid approach, while speed is close to the features scaling approach.

While favoring detection speed from all points of view is crucial (Sect. 4.5), the computational burden is not the only problem introduced by the need of detecting pedestrians at multiple distances. Farther away pedestrians correspond to less pixels in the image, thus, less details are captured than for closer ones. Additionally, due to the depth of view of the cameras, different degree of sharpness is observed for different distances. According to the usual camera settings for ADAS applications, farther away pedestrians are more blurred. In summary, pedestrian imaged appearance changes according to the distance (Fig. 3.3). This implies that a single resolution pedestrian model may be too blurred to work at different resolutions, which can affect overall detection performance. In fact, this situation happens for almost any target to detect in the ADAS context (vehicles, traffic signs, etc.). According to our experience in this field, using different models for different resolutions turns out in better accuracy. In [193], we follow such approach for detecting cars at nighttime to control an adaptive headlamp system in a range of distances that runs from a few meters to half a kilometer.

Accordingly, in pedestrian detection the idea of using resolution-dependent models has also been explored. Note that the ideal case would be to have a different model per resolution (Fig. 3.3). Of course, this can be a naive approach since many models would have to be trained if we need to use many resolutions, and depending on the training process and the available resources (training data, computation power, etc.) this could be overwhelming. Moreover, models for close resolutions would be too similar, which would not be exploited by such a naive procedure.

Hence, different researches have used resolution-dependent pedestrians models, but not one model per resolution. For instance, in the previously mentioned work by Hou et al. [144], it is not trained a single multi-view pedestrian classifier but three of them are trained to gain robustness with respect to multi-resolution. In particular, the sizes of the respective canonical windows are (in pixels) 24×58, 36×87 and 48×116. It is worth noting that there is not a canonical window standard used by all researchers since this also depends on the testing data set, especially on the median pedestrian height (Table 5.1).

Analogously, Goto et al. [126] also train a multi-view (back, front, left and right views) three-stage pedestrian classifier. First and second stages are based on HOG features without block overlapping and with it, respectively, and are intended to provide fast rejection of non-pedestrian candidates. The third stage is based on FIND features (Fig. 3.2, co-occurrences). In all stages linear SVM is used as classifier. Multi-view analysis is only considered at the third stage, whereas three resolutions are considered at all stages, namely for near (0–20 m), middle (20–40 m) and far (40–60 m) pedestrians. Classifiers are trained using view-resolution specific pedestrian examples. The size of the candidate windows determines the resolution-dependent classifier to apply at any stage.

In [116], Ge et al. use a two stage pedestrian classifier for nighttime pedestrian detection in near infrared images (NIR). The first stage is based on Haar-like features and Gentle AdaBoost. This stage is used for rough but fast candidates rejection. The second stage applies HOG-based classification also using Gentle AdaBoost, each weak classifier depending on each component of final HOG feature vector. In fact, this second stage relies on three resolution-dependent HOG/AdaBoost classifiers, i.e., according to the original size of the candidate window it is applied the classifier for small, medium or large distance pedestrians. In both stages, each AdaBoost weak classifier is based on a classification/regression tree operating on a specific feature.

A remarkable approach dealing with multi-resolution is the so-called *Very Fast* by Benenson et al. [18]. It is similar in spirit to the FPDW of Dollár et al., even relying on the integral channel features, but rather than doing a sparse pyramid of the input image at detection time and using features approximation for the in-between resolutions, Benenson et al. translate this idea to training time. In particular, a sort of sparse pyramid of resolution-dependent classifiers is trained. During detection time, the range of resolutions to be explored is covered by these classifiers and proper approximations of them are used for the resolutions that they do not explicitly cover (Fig. 3.3). In short, since integral channel features are used to learn a strong classifier from a set of decision trees, each tree is based on three decision stumps. Accordingly, the re-scaling of the explicitly learned classifiers is based on re-scaling the rectangles of the involved integral channel features and the thresholds of the decision stumps. In this case, during training all the available examples are re-scaled to all the canonical window sizes of the learned classifiers, i.e., those of the explicitly learned resolutions.

Not only holistic models and multi-view ones have been extended to deal with multi-resolution. The former deformable parts-based model based on HOG/Latent-SVM by Felzenszwalb et al. [93] has been also extended in this sense by Park et al. [243]. The proposed method acts as a deformable parts-based model when classifying large-size candidate windows and as a holistic template when classifying small-size candidate windows, i.e., when appearance details are missing due to distance to the camera, parts detection is not required. Notably, during training, clearly large and small resolution pedestrians are considered as belonging to such large/small resolution classes. Then, for intermediate resolution pedestrians, resolution is considered as a latent variable which is estimated through the latent SVM framework.

3.3.6 Occlusion Handling

The works revised so far in this chapter mainly focus on obtaining pedestrian classifiers robust to inherent pedestrian intra-class variability (height, pose, clothes), imaging conditions (camera viewpoint, resolution), and environmental factors (illumination, background). However, pedestrians do not walk alone: there are more active traffic participants, e.g., vehicles and other pedestrians. The immediate consequence is that partial occlusions happen among them. This must be explicitly taken into account by pedestrian classifiers to avoid a severe drop in accuracy. Additionally, segmenting which pixels of the detected pedestrians are occluded can help the further tracking module to improve the appearance dynamic model of the pedestrians to track, i.e., by only using non-occluded pedestrian pixels.

Traditionally it is argued that parts-based methods can handle occlusions because the miss detection of one part does not affect the detection of the rest of parts. One can use a similar argument for holistic models based on grids of local features, since the occlusion of a grid element does not necessarily affect the results on the other grid elements. However, results presented by Dollár et al. [71] reveal that if occlusions are not explicitly taken into account, even non-heavy partial occlusions give rise to a very poor accuracy. Arguably, this may be due to the fact that local evidences (either parts or local features) are jointly taken into account to compute the final pedestrian/non-pedestrian classification decision. Hence, how to fuse these evidences in presence of partial occlusion has to be explicitly designed or learned, in other words, good results under partial occlusion do not come *per se*.

PPSs need to detect pedestrians as soon as possible to track their trajectory and take risk-based decisions with low latency. For instance, a pedestrian can suddenly appear behind a parked car, thus, he/she appears into the scene under partial occlusion (Fig. 3.1) but the PPS should ideally detect him/her at that moment. According to the relevance of this issue and the poor results reported, when we presented [121] one of our recommendations was to focus part of the research effort on these problematic cases. In line with this, during the last years several works have addressed this topic as we briefly summarize to conclude this section.

In fact, we have already mentioned a work in which partial occlusion is explicitly handled. We refer to the work of Wang et al. [315] where HOG+LBP features are used to build a holistic pedestrian classifier. Basically, it is done an analysis of HOG blocks to determine if they depict a pedestrian local patch or not. According to the blocks classified as not depicting pedestrian patches, it is decided if there is an occlusion or not. Occlusions are classified as upper or lower body ones. If it is determined that there is an upper body occlusion then a lower body part classifier is run, whereas if there is a lower body occlusion then an upper body part classifier is applied. Hence, in case of occlusion a weighted summation of the holistic classifier and the parts-based classifier is used to take the final pedestrian/non-pedestrian classification decision.

Tosato et al. [294] adapt pedestrian classification on Riemanian manifolds to deal with occlusions. In particular, the presence of four types of occlusions are tested (top, bottom, left, right). Evidence of such occlusions is collected by the weak classifier involved in the pedestrian classification process provided they operate on local patches. In this case, however, the aim is not to provide a pedestrian classifier more robust to occlusions. The occlusion reasoning is applied to already detected pedestrians. The detected occlusions are used then as additional information for further processes (e.g., tracking).

The previously commented mixture-of-experts framework defined by Enzweiler and Gavrila [85] has been also evaluated for handling partial occlusions by Enzweiler et al. [81]. In this case the experts are per modality (luminance, stereo depth, optical flow motion) and per body-inspired parts (head-shoulders, torso-arms, waist-legs). In particular, each expert is a classifier based on linear SVM and HOG features extracted from the corresponding modality and part. Occluded pixels are determined using segmentation in the depth and/or motion modalities as well as measuring the similarity of the resulting segments with pre-defined multi-view pedestrian shape masks (e.g., see Fig. 3.5). Such similarity is evaluated for each part individually. Overall, the more occluded pixels are determined to belong to a part, the less weight receives the corresponding experts when computing the final score of the mixture-of-experts classifier.

In [323], Wojek et al. use a pedestrian model of seven components, namely a HOG/Latent-SVM parts-based classifier [92], and six body-inspired part classifiers based on HOG and histogram intersection kernel SVM (HIK-SVM by Maji et al. [197]). The former is assumed to properly detect fully visible pedestrians while the body-inspired part classifiers act as a mixture-of-experts whose individual weight depends on the visibility of the image area in which they operate. Such visibility is inferred through a probabilistic framework for spatio-temporal 3D scene analysis based on a monocular acquisition system with odometry.

Interestingly, occlusions have been also explicitly modeled trough the concept of grammar. For instance, Girshick et al. [125] do it for the HOG/Latent-SVM case by incorporating the *occluder* part into the model, which captures the appearance of the typical stuff occluding people. Duan et al. [75] take the organization of text grammars (i.e., words-sentences-paragraphs hierarchy) to define pedestrian models. Basically, words correspond to sub-windows of the canonical windows, sentences are combination of blocks and paragraphs are combinations of sentences. Multiple-aspects (eight) and different occlusion types (five) are modeled as different paragraphs. Real-Adaboost mining APCF features is used at the three levels of the grammar.

Since many times pedestrians are occluded by other pedestrians, some approaches takes this fact explicitly into account to gain robustness to occlusion. For instance, this is the idea behind the work of Tang et al. [289], who take advantage of the characteristic appearance patterns occurring in person-to-person occlusions.

Finally, we would like to mention also the work by Ouyang and Wang [235] who propose a deep model to learn the visibility relationships among parts, i.e., with the purpose of ensembling the results of the different part detectors more accurately.

3.4 Training

Along Sects. 3.2 and 3.3 we have been constantly mentioning the training of classifiers using different learning frameworks, namely, SVM, AdaBoost, NN, ECOCs, etc. In order to perform such a training, positive samples (pedestrians) and negative ones (background) are required. They come from datasets (e.g., Table 5.1) with annotated ground truth (e.g., Fig. 5.1). Pedestrians are annotated at least with a bounding box in a set of *positive images*. Not all pedestrians need to be annotated within the positive images, therefore, for randomly collecting background samples easily, a set of *pedestrian-free images* is also provided. In any case, all samples, pedestrians and background, must be standardized to a canonical window size (or several for some multi-resolution approaches) during learning (Fig. 3.23). The training data and several good practices are critic for obtaining the best accuracy possible from the features and learners at hand. We focus on some of these aspect in the rest of this section.

3.4.1 Parameters Tuning

One of such good practices is parameter tuning by using either a validation set or cross-validation [172], although the latter can be too time consuming. The same training set with the same features and learning algorithm can give rise to very different classification accuracy depending on the parameters' value. This is not an easy task since features and learning algorithms have their own parameters. For instance, popular HOG/Linear-SVM approach requires to search the best value for many parameters [60] (HOG: cell size, block size, histogram bins, block overlapping, etc.; Linear SVM: C parameter). Even not being an easy task it can pay back by producing a significant improvement in the accuracy of the learned classifier. Of course, parameters optimization is a research branch in itself out of the scope of this book, but we think it is worth including this small reminder here.

3.4.2 Bootstrapping

Another good practice is the so-called *bootstrapping* mechanism. Most common bootstrapping consists of two steps. First, the current version of the pedestrian detector is applied to the training pedestrian-free images for easily collecting false detections, i.e., the so-called *hard negatives*. Second, the background samples used so far for training the current pedestrian classifier are augmented with these new hard negatives, and the pedestrian classifier is retrained. These two steps can be iterated. Note that the *background class* is quite heterogeneous since it corresponds to anything on outdoor scenarios out of pedestrians. Therefore, bootstrapping allows to progressively diversify the background samples available for the final training.

Learning a classifier without a proper bootstrapping can severely drop the accuracy that otherwise can be potentially achievable given the overall training set and learner in use. Most of the cited works so far mention that bootstrapping is done. What can vary from work to work are the details of how to do it since there is not a general optimum recipe. For instance, collecting the initial background examples is the first issue. Then, during bootstrapping it is possible to limit the number of hard negatives to collect per pedestrian-free image, or just collect all them. Moreover, the hard

Fig. 3.23 Training samples (pedestrians and background) from different datasets (Table. 5.1), standardized to a canonical window size. In the case of the real-world images (Daimler and INRIA) the pedestrian bounding boxes are provided manually. In the case of the virtual-world images the bounding boxes are automatically obtained by accessing video game information [199, 200]. Silhouettes can be easily obtained from virtual data too

negatives can be collected before or after applying the non-maximum-suppression step of the classifier (Sect. 4.2). Additionally, if the pedestrian-free images are, in fact, video frames, then hard negatives from consecutive frames can be too similar, i.e., the bootstrapping can be collecting *repeated* hard negatives. Finally, as we have mentioned so far, current negative samples are augmented with new hard ones at each bootstrapping round, but another possibility is to remove the current negatives that are not hard anymore and, thus, keeping a smaller cache of negative samples at each round [92].

It is also worth noting that in testing/operation time of a pedestrian detector some of the false positives can be due to body parts. Using pedestrian-free images to collect background does not bring such cases as hard negatives. However, provided that training pedestrian are annotated with bounding boxes, body parts can be used as negative samples for training full body classifiers [325]. Additionally, if the candidates generation method (Chap. 2) is not a sliding window variant but it is based on image content cues (depth, motion, etc.), then negative samples, also hard ones, a priori should be collected from such type of candidates to lead the classifier learner to focus on the specific background that must actually be distinguished from pedestrians (e.g., see the dataset used in [81, 85]; it was also the approach we follow in [120, 122, 123]). In fact, if we think of such advanced candidates generation as a first stage of the pedestrian classifier, this is just a particular case of bootstrapping during the learning of a cascade of classifiers. In these cases, bootstrapping can be used for conditioning the negatives of a given stage/layer by the hard negatives of the previous one. For instance, this is the case of AdaBoost-like methods that build the strong classifier as a rejection-cascade [309].

Yet a very important question is when to stop the bootstrapping [220, 312]. Some authors fix a maximum number [60, 72, 93]. Other stopping criteria are saturation of a classification accuracy measure evaluated for the training set, low percentage of new hard negatives, or saturation of the training resources (processing time and/or computer memory). In case of using a validation set, stopping can be based on the saturation of a classification accuracy measure evaluated for such set.

We have tested different methodologies along the years and currently we are confident by starting with an equal number of pedestrians and background samples for training the first classifier version, and then lead the bootstrapping iterations to collect most of the background samples from which the final classifier is learned. We collect the initial background samples randomly but visiting the different pedestrian-free training images, with a frame step for videos. We usually do not put any limitation per pedestrian-free image during bootstrapping, but we use a frame step for videos. From one bootstrapping run to another we can force to process different frames of the video. We collect the false positives before the non-maximum suppression step of detection. The particular settings of the procedure depend on the processing time to learn each new version of the classifier as well as the computer memory available to fit the pedestrian and background training samples. As stopping criteria we either use saturation of the average miss rate in per-image evaluation (Sect. 5.2) provided we have a validation set, or low percentage of new hard negatives in the bootstrapping round.

3.4.3 Data Annotation

Pedestrian and background samples come from manual annotation, namely, at least the pedestrians' bounding box is delineated and a *yes/no*-decision to identify pedestrian-free images is provided. Collecting good samples of pedestrians and background is very important for training since with poor samples even the best combination of features and learning machine cannot provide a good pedestrian classifier. For instance, let us assume we want to learn a pedestrian model based on HOG/Linear-SVM. If the different pedestrian samples are poorly aligned within the provided bounding box, the learned pedestrian model can be too blurry provided miss-alignment exceeds HOG cells size. We note that bounding box annotation is not the most complex task, for instance, the chamfer system [110, 112] and aspect-dependent HOG features [189–191] use manually annotated pedestrian silhouettes. Some parts-based methods require manual annotation of parts information [36], or even when parts are automatically searched the result can be sensitive to the heuristic initialization [92, 93]. In fact, the annotation of silhouettes can be also alleviated by applying motion segmentation to videos taken from a static camera [273] or performing appearance assisted segmentation [252].

Since manual annotation is a tiresome process prone to errors and imprecision, different methods have been proposed to alleviate it. We can distinguish two tendencies, namely, developing new annotation paradigms and engineering samples. A new annotation paradigm consists in the use of web-based tools. A well known example is LabelMe [265] which allows human *volunteers* to localize image objects of a established class category by framing them with polygons. Nowadays, however, Amazon's Mechanical Turk (MTurk) [258] probably centralizes the most powerful web-based annotation force. MTurk allows researchers to define *human intelligence tasks* (HITs: *what* and *how*) of different difficulty (e.g., from marking points of interest to drawing polygons) to be taken by human online workers (*turkers*) which are paid for their work. Thousands of annotations can be collected through MTurk. Unfortunately, as it is argued by Berg et al. [19], where such web-based tools are analyzed, *it is a fallacy to believe that, because good datasets are big, then big datasets are good*. A key reason behind such a fallacy is the human factor. For instance, most turkers are not computer vision experts and have not scientific motivation, which makes annotation quality very sensitive to both annotation instructions [80] and economic reguard [19]. In fact, due to such reasons and even to malicious workers, it is necessary to collect multiple annotations from the same image/object and assessing annotation quality from them [19]. Thus, *how to collect data on the internet* is a non-trivial question that opens a new research area [19] involving *ethical issues* [264] since human paid work is at the core.

Aiming at collecting useful pedestrian samples and reducing manual annotations, the so-called *active learning* [54] has also been explored by Abramson and Freund [1] as a new annotation paradigm. Active learning basically consists of a stage to obtain an initial classifier, followed by a loop in which the current classifier is plugged in a detector that is applied to unseen images, a *human oracle* performs *selective sampling* (i.e., he/she annotates image windows that fall into the classifier *ambiguity region*) and then previous and new annotations are used for re-training the classifier. Active learning is especially interesting to collect informative pedestrians since, as we have seen, hard negatives are satisfactorily collected by bootstrapping.

In these proposals, positive samples are annotated at existing images. Alternatively, we can *engineer* them as Enzweiler and Gavrila [82], who synthesized pedestrians by transforming their original shape, texture, and surrounding background. Aiming at increasing variability, such a transform is applied according to selective sampling within an active learning framework. However, for synthesizing the pedestrians manual silhouette delineation is required. In [253], samples with silhouette delineation are obtained automatically by Pishchulin et al. through elaborated 3D human synthetic models. Using such synthetic samples, it is obtained an accuracy close to the one based on real-world samples. Besides, direct fusion of all synthetic and real-world samples gives rise to the best pedestrian classifier for the evaluated pedestrian descriptors and learning machines. However, for embedding the synthetic pedestrians in plausible poses on the ground plane, camera-calibrated human-free driving sequences must be recorded. In fact, obtaining manipulated good-looking images of people by performing holistic human body transformations is in itself an area of research, specially when video is involved and thus temporal coherence is required [159].

Marín et al. [200] proposed photo-realistic virtual worlds (Virtual dataset in Table 5.1) for collecting training samples. Following such an approach detailed ground truth is automatically available for each virtual-world pedestrian, i.e., its bounding box, silhouettes and parts. Pedestrian-free images are automatically generated as well. Yet, the challenge consists in achieving good pedestrian detection performance with real-world images using classifiers learned from such virtual-world samples. Indeed, Marín et al. show that using a virtual city to collect the training samples and real-world images of urban scenarios for testing (Daimler-Detection dataset, see Table 5.1), the obtained accuracy is equivalent to the one achieved by training with real-world manually annotated samples coming from urban scenarios.

Indeed we think that virtual worlds are an interesting framework to explore since it is not only that some modern video games, life simulators and animation films, are gaining photo-realism, but also the whole simulation pyramid for creating an artificial life is being considered: visual appearance both global (3D shape, pose) and local (texture, in which involved Computer Graphics aim at approaching the power spectrum of real images [255]), kinematics, perception, behavior and cognition. For instance, see [275] for the case of animating autonomous pedestrians in crowded scenarios. This means that from such virtual worlds we could collect an enormous amount of automatically annotated information in a controlled manner.

Yet another consideration is the relationship between the diversity of pedestrian examples required for training and the used image descriptors and pedestrian models. Basically, we can say that the more invariances achieved by the joint action of descriptors and models the less pedestrian examples are required. For instance, HOG is robust to shifts withing the cells size, therefore, jittered pedestrian samples are not required. Probably, this is not the case for classifiers based on NNs where no predefined features are provided, thus, invariance to jitter must be learned. Analogously, given a typi-cal pedestrian view (e.g., side view) a deformable parts-based model ideally would require less examples of pedestrian poses than a multi-pose holistic model, in other words, the deformable parts-based model implicitly encodes unseen poses up to certain extent.

3.4.4 Domain Adaptation

Finally, we think it is worth to mention an additional issue regarding classifiers training. Let us imaging the following situation. We annotate our training set of pedestrians using images acquired with a given camera, and learn a pedestrian clas-sifier that solves our application. Say that later we shall use a different camera or we have to apply the classifier in another similar application/context but not equal. This variation can decrease the accuracy of our classifier because the probability distribu-tion of the training data can be now much different than before with respect to the new testing data. This problem is referred to as *dataset shift* and is receiving increasing attention in the Machine Learning field [260] due to its relevance in areas like natural language processing, speech processing, and brain-computer interfaces, to mention a few. In fact, this problem has been reported by Vázquez et al. [308] in pedestrian detection and by Torralba and Efros [292] for object detectors. Finding solutions to the dataset shift problem is the objective of *domain adaptation* techniques.

Vázquez et al. address this problem for pedestrian detection when the training set comes from virtual-world data and, of course, the testing is performed in real-world images (Fig. 3.24). On the one hand, Vázquez et al. propose domain adaptation based

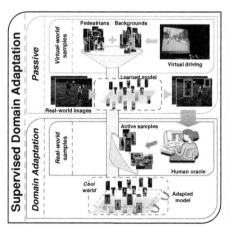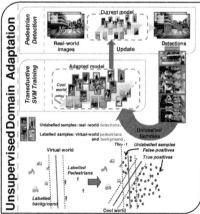

Fig. 3.24 Domain adaptation (DA) between virtual- and real- world data. *Left* supervised DA based on active learning. *Right* unsupervised DA based on self-detections and transductive-SVM

on active learning for holistic pedestrian detectors relying on both HOG/Linear-SVM [307, 308] and Haar+EOH/Real-AdaBoost [306]. Basically, by fusing the virtual-world data with relatively few actively collected real-world data, the domain shift problem is removed. On the other hand, Vázquez et al. have also tested an unsupervised domain adaptation technique, i.e., not requiring manual annotation of real-world pedestrians, based on Transductive-SVM [305].

3.5 Discussion

This chapter is by far the largest of the book, which correlates with the major research effort so far. In other words, finding the best possible pedestrian classifier has received the main attention among the different pedestrian detector modules (Fig. 1.4). Then a natural question that arises is *which is the best pedestrian classifier?* However, it is very difficult to answer this question. In fact, we have avoided this point of view along the chapter because for a long time there were not standard evaluation protocols and datasets. Moreover, different implementation details can significantly boost the classification accuracy based on a given feature and the same happens by following a proper parameters optimization (using a validation set or cross-validation) of the features and learners, but this crucial issues not always have been well clarified in the literature. Of course, nowadays this circumstance is being changed (Chap. 5) and light has been put on comparing different pedestrian detectors [74], i.e., at least one can identify groups of best approaches that seem to perform similarly good. Interestingly, there is a web page,[2] at the moment maintained by Dollár and Wojek, centralizing evaluation efforts according to the proposal in [74]. For instance, in terms of detection performance it seems that the multi-resolution proposal by Park et al. [243] is the current state-of-the-art. Just to be able to check this information is a great advance reached since we did our first survey work [121], where to have such protocols was one of the *to-do* items we pointed as really necessary for the field. Of course, in addition to [74], we must credit also other comparative studies, for instance [83, 85, 152, 220, 322] just to mention a few.

In spite of that, it is still difficult to say which is the best singe feature (HOG, LBP, etc.), or which is the best pedestrian model (holistic, multi-aspect, parts-based, etc.), since the classifier is not the only stage affecting detection performance. For instance, the detector mentioned before as best performing one [243] only uses HOG features, but a multi-resolution deformable parts-based pedestrian model learned by a Latent-SVM framework. However, not too far away in detection performance we find the *crosstalk* detector [74], which is based on the integral channel (combination of) features and AdaBoost but follows an holistic approach (i.e., neither deformable parts are used, nor multi-aspect, nor multi-resolution) and tightly couples detector evaluation of nearby candidate windows.

---

[2] www.vision.caltech.edu/Image_Datasets/CaltechPedestrians/

Even the difficulty to point out the best single type of feature and pedestrians model from the detection performance viewpoint, there is a clear conclusion from the reviewed literature, namely, *diversifying* the captured information boost the detection performance. It can be done by either combining image descriptors (e.g., HOG+LBP, Haar+EOH, integral channels, etc.), or classifiers ensembles, or multiple modalities (e.g., HOG from luminance, depth, and motion), or multiple aspects pedestrian models, or multiple parts models, or multiple resolutions ones, or combinations of several of these possibilities. Regarding such diversification, there are some trends that we think worth to be highlighted:

- *Multiple modalities.* Different physical modalities (luminance/color, depth, motion) really convey complementary information able to boost the pedestrian detection performance [74, 85, 312, 325]. Moreover, cues as depth and motion can be used to reduce the number of candidate windows to process as well as to take high level decisions according to the distance and motion of the detected pedestrians with respect to the ego vehicle. Diversity thanks to multi-modality seems to be superior to diversity due to different image descriptors.

- *Multiple resolution models.* The same pedestrian imaged from different distances to the camera looks different in the corresponding images, even if his/her aspect and the external factors (lighting, occlusion) do not change. This is due to the camera resolution as well as the depth of focus. Thus, trying to convey a large range of distances into a single resolution pedestrian model (i.e., a single canonical window) a priori will turn out in a lower detection performance than by using some strategy to learn different pedestrian models for different resolutions [18, 273]. Note that, for instance, in [243] it is reported that a holistic model works better for small pedestrians than a parts-based one, and the opposite for large pedestrians, where small and large refer to image pixels.

- *Implicitly modeling unseen examples.* Provided the difficulty of collecting a comprehensive set of pedestrian examples, if the pedestrian model is able to implicitly codify unseen examples, a priori the resulting template/classifier will allow better detection performance. For instance, given a fixed pedestrian view, implicit shape models [182, 272, 273] and deformable parts-based models [92] have such a potential ability. In the same sense, if a feature is robust (invariant) to certain variability conditions we would not need all kind of pedestrian examples showing such variants to learn the pedestrian classifier. For instance, by using unsigned gradient directions in HOG features, we are simultaneously modeling both light-clothed pedestrians on a dark background and the opposite, without needing both types of pedestrian examples. In this sense, not manually engineered features a priori may need more examples to learn a classifier since they need to learn the invariances. On the other hand, manually engineered features are limited by the knowledge of their human designers.

- *Sub-categorizing the data.* Focusing on the deformable parts-based model, the previous intuitive affirmation seems to be controversial. Zhu et al. [343] support the fact that supervised deformation part models are able to capture unseen information. However, HOG/Latent-SVM scheme [92] does not exhibits such an ability (lacks supervision). Divvala et al. [68] also reports that the parts' deformations can be totally removed from the model provided a proper aspect-driven sub-categorization of the training data is performed and a multiple-aspect detector is build. The use of good strategies to sub-categorize the data is also highlighted as crucial by Zhu et al. [343]. Moreover, it seems that using appearance-based categorization is at the same level than using categories manually annotated [68]. Interestingly, Parikh and Zitnick [242] show that in the HOG/Latent-SVM model the weakest component is parts detection, not the parts deformation penalty. This study is based on a single aspect, thus, parts filters are probably more blurred that it would be by using a multiple aspect approach combined with the multiple parts. Note that HOG/Latent-SVM allows for a rought separation of the training data in two aspects, however, it seems too poor [68] and, in fact, not used in [242].

- *Training algorithm.* SVM and AdaBoost variants are the dominant algorithms in pedestrian detection. Looking at the literature it is difficult to say if there is a *winner*. In fact, Zhu et al. [343] claim for better features not better training algorithms. However, regarding the computation time, rejection cascades based on AdaBoost variants seem to be a better choice since they can avoid the computation of many features at background areas of the image, which are the most frequent. This is specially important when the available features form a very high dimensional space.

Looking at these highlighted points, it is interesting to see that researchers of two major industrial actors of the pedestrian detection field, namely, Daimler AG and MobilEye, have presented research papers in agreement with some of the above highlighted points. For instance, on the one hand, as we have mentioned in Sect. 3.3 (multiple aspects), Enzweiler and Gavrila [85] propose a mixture-of-experts scheme that uses multi-modality (luminance, dense stereo, horizontal optical flow), multiple complementary descriptors (HOG, LBP), and multiple aspect weighting of the training pedestrians based on shape cues (chamfer system or the like). On the other hand, as we have seen in Sect. 3.3 (multiple parts), Shashua et al. [276] use a model of non-deformable multi-level overlapping parts based on a sort of simplified HOG features, and the training is done according to a clusterization of the data based on aspect and illumination conditions. This clusterization is reported as crucial by the authors. Notably, in both works [85, 276] it is reported that, during training, background samples do not come at random but from the candidates generation module, thus, pedestrian/non-pedestrian discriminative classifiers focus directly on difficult cases. We used also this approach in [120, 122, 123].

As we see, blindly increasing training data size and pedestrian model complexity do not help to build pedestrian classifiers of higher accuracy. Instead, having diversified training data and being able to follow a cluster-dependent training is a major key. Under this assumption we are personally focusing on the use of synthetically generated scenarios [199, 200] since we automatically can have all kind of associated pedestrian information (appearance, silhouette, parts, aspect, occlusion degree, etc.). However, training with images taken with one camera and testing on images acquired with a different camera can cause dataset shift. Thus, training with virtual-world images to operate on real-world ones can suffer domain shift. Because of that domain adaptation techniques are also quite relevant for us [305–308]. In particular, we think that on-line domain adaptation can boost pedestrian detection performance.

We conclude this chapter by remarking that pedestrian detection under partial occlusion is a very relevant issue still open.

Chapter 4
Completing the System

Although most of the responsibility of a PPS lays in the candidates generation and classification modules, other additional components are needed to fulfill the system requirements and complete the process. They are the preprocessing module, which is placed at the beginning of the system's pipeline before the candidates generation; the verification and refinement, which are placed just after the classification; and the tracking and application modules, which are placed at the end of the system's pipeline and it is the one that generates the output. In addition to reviewing such modules, in this chapter we also review the approaches employed to achieve real-time, which is a global aspect to take into account in a real system.

4.1 Preprocessing

In this module we describe a set of techniques focused on preparing the image for further processing and extracting relevant information for later modules. These techniques, which are detailed in few papers, may be divided into three groups: input image adjustments, camera calibration and modality acquisition.

Low-level adjustments such as exposure or dynamic range are usually not described in PPSs literature given that they are not the research focus in this area. Nevertheless, they can represent a recurring difficulty, especially in urban scenarios. Tunnels and narrow streets may result in over/under saturated areas in the image or poorly adjusted dynamic range, which creates additional difficulties for the latter algorithms of the system. In order to tackle these problems, but not specifially addressed at ADAS, Nayar et al. [224] present some approaches for performing a locally adaptive dynamic range: fusion of different exposures, spatial filter mosaicing and pixel exposures, multiple image/pixel sensors, etc. Besides, during last years, solutions using *high dynamic range* (HDR) images [169, 202, 304] are gaining terrain in driver assistance because of their potential to provide high contrast in the aforementioned scenarios.

D. Gerónimo and A. M. López, *Vision-based Pedestrian Protection Systems for Intelligent Vehicles*, SpringerBriefs in Computer Science, DOI: 10.1007/978-1-4614-7987-1_4, © The Author(s) 2014

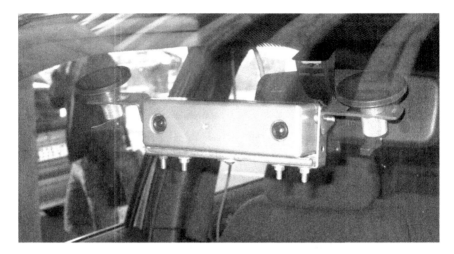

Fig. 4.1 Prototype of a stereo acquisition system with a commercial Bumblebee stereo rig

Camera calibration is needed to compute the correspondence between the 3D world and the 2D image. Figure 4.1 illustrates a prototype of a stereo acquisition system. Some approaches tackling intrinsic and extrinsic on-board self-calibration are presented in [38, 62]. It is important to note that the calibration step is normally made once and then assumed fixed. Nevertheless, if the camera position with respect to the road is altered some algorithms such as 3D reconstruction will work incorrectly with catastrophic results for the system performance. For this reason, continuous stereo self-calibration algorithms are being proposed. In [63], the authors propose a continuous recursive auto-calibration algorithm based on an Extended Kalman Filter of the parameters. In [301], Unger et al. propose to increase the search range for correspondences, which overcomes the problems of inaccurately rectified images.

Another important part of preprocessing corresponds to the image modality that is used. Different modalities (and in some cases a combination of them) are used to detect pedestrians: *temperature, luminance/color, depth, motion flow*, etc. (Fig. 4.2). The way these images are acquired and processed is also important. For example, depth is computed from stereo rigs, which are mainly based on luminance/color cameras, thus having vision-based depth information implies to have luminance/color information available. Motion flow is mainly computed from luminance/color cameras as well, therefore the motion information comes together with luminance/color. When we compute the motion flow from consecutive monocular images we term it *optical flow*, while if it is computed from consecutive depth images we call it *scene flow*. Scene flow corresponds to real-world motion, while optical flow is confined to image-world. The depth image is affected by the camera calibration previously mentioned, and the luminance image depends on the aforementioned low-level adjustments. If some of the components is incorrectly adjusted, the depending modalities will be affected. Furthermore, color is obtained from a single imaging

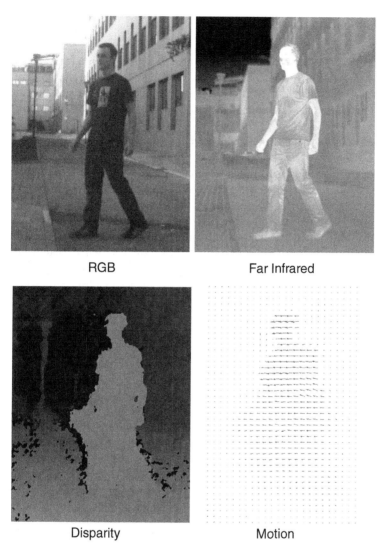

RGB   Far Infrared

Disparity   Motion

Fig. 4.2 Different modalities used in PPS: the information which is fed to the system algorithms significantly varies depending on the sensor

sensor based on a Bayer pattern, thus, they are usually less sensitive and of worse *actual* resolution than their counterpart luminance sensors (i.e., so-called mono-chrome sensors). Therefore, although we can compute luminance information from Bayer-pattern color images, in general they are of worse quality than the equivalent obtained from a monochrome sensor. This raises an important issue to take into account when designing a PPS: is color information a key component for the system? If it is not, monochrome cameras can provide higher quality images.

4.2 Verification and Refinement

The expected classification output is a set of redundant windows (i.e., overlapping the same target) mostly containing true positives but also some false positive. The objective of this module is to process these classified windows and output only meaningful detections (one detection per target) filtering out the false positives. The process is divided into two steps: verification, which discards the remaining non-pedestrians; and refinement, which groups the overlapping windows and even provides a fine pixel-level segmentation.

Although the performance rates of modern pedestrian classifiers is extremely high, it is still acceptable to get some intermittent false positives from frame to frame. In order to filter out these undesirable detections a common approach is to *re-classify* all the windows with a criterion not overlapped with the classifier. Gavrila et al. [113, 114] verify detections by performing cross-correlation between the left image of a stereo pair and the isolated silhouette computed by the Chamfer system in the right image. In [99], Franke and Gavrila suggest the analysis of gait pattern of pedestrians crossing perpendicular to the camera. The target must be tracked before applying this method, thus the order of verification/refinement and tracking modules is interchanged for this particular technique. In [276], Shashua et al. propose a multi-frame approval process that consists of validating the pedestrian-classified windows by collecting information from several frames: gait pattern, inward motion, confidence of the single-frame classification, etc. In this case, verification follows tracking. Ramanan [262] filters out false positives with a figure-ground segmentation using graph cuts and color histograms. Finally, in [185], the authors use chamfer matching to both verify and refine the found pedestrian shapes in an explicit verification step.

Once the false positives have been filtered out, the refinement stage converts the raw windows into detections. The typical behavior of a classifier is to not only provide a peak at the correct target's position and scale but also output weaker responses around it. Accordingly, a common approach is to transform the overlapping responses framing a pedestrian in a single detection by grouping them with some clustering algorithm. This process is known as non-maximum suppression (NMS). Although the need of NMS was already present in the first pedestrian detection papers [241, 310], it was not until the publication of [59] that this process was detailed. In [59], Dalal makes use of mean-shift clustering [55] to find the minimum set of windows that best adjust to the positive windows distribution. First, the windows are projected into a x-y-scale space, where the overlapping ones are grouped in the same region, and then mean-shift iteratively searches the modes of the windows in this space. For the sake of completeness, it is worth mentioning the work by Agarwal et al. [2]. Their proposal, which was tested for vehicle detection consists of two algorithms. The first creates an activation map, where high-confidence detections mark their neighborhoods as invalid for new detections. Given that this system is based on a parts-based classifier, the second algorithm constrains the parts to be assigned to only one detection, and thus non-maximum detections are discarded by iteratively decreasing their confidence. Several authors such as Laptev [175] and

Fig. 4.3 Accumulative clustering efficiently groups overlapping detections framing single pedestrians (figure reproduced from [119])

Gerónimo et al. [119] propose faster clustering algorithms by directly comparing the overlapping area between windows (see Fig. 4.3). In both cases, the degree of overlapping between the windows is compared by using the PASCAL criterion [195]: $A \cap B / A \cup B$, where A and B are windows or cluster representative, depending on the algorithm. Following the same pairwise overlap comparison idea, Dollár et al. [72] state that the overlap criterion $A \cap B / min(A, B)$ provides better performance. It has been proven that mean-shift is slower than pairwise algorithms, while these latter can provide the same performance [119].

In a recent work, Gressman et al. [129] propose two different approaches to refine the detections' windows (in the sense of perfectly framing the given pedestrian) that can be run after the clustering step. Both approaches use neural networks with LRFs. The first approach uses a trained ranking classifier that determines the order of detection windows in regard to their coverage of the pedestrian. In the second a binary classifier is trained to stepwise move an initial detection's window towards the optimal position.

Several techniques have been proposed to not only output bounding box detections but also the pedestrian's silhouette. In [39], by Broggi et al., the head-and-shoulders' silhouette that is matched during classification is taken as a reference for refining detection down to the feet, by using the vertical edges computed for the symmetry detection. The accurate location of the feet is then used to compute the distance to pedestrians by assuming a planar road. Then, a stereo processing completes the refinement by correlating the left-image bounding box to certain positions of the right image. Caballero et al. [46] proposes a bag-of-words representation of segment features to output the rough silhouette of the pedestrian. Andriluka et al. [5] estimate the 3D pose of pedestrians by using 2D limb tracklets in a tracking-by-detection framework. Horbert et al. [143] propose two different models to perform segmentation based tracking: the first using hierarchical appearance models and the second using shape information from classification.

Some techniques that use FIR images are 2D model matching [22]; 3D model matching [25, 42]; symmetry [24]; and a multiple filter approach based on the overlapping area between positively classified ROIs and group multiple windows in a single detection [196].

Finally, it is worth mentioning that some classification algorithms such as the Chamfer System by Gavrila et al. [113] or the ISM [182] (see Chap. 3) are capable of providing shape-based detections without extra-refinement steps. In fact, they have been combined in a single verification-refinement process in [185].

4.3 Tracking

The tracking module follows the detections (classified, verified and refined windows) over time. This step has several purposes: avoiding false detections over time, predicting future pedestrian positions (which may be fed to the candidates generation module) and, at a higher-level, making useful inferences of the pedestrian behavior as walking direction or time to collision. Traditionally, tracking has received less attention than other modules such as classification. The most common algorithms used are Kalman and Particle Filters, though other interesting approaches are also described in this section.

Kalman filter is the most extended approach. Researchers from Daimler AG proposed the use of two Kalman filters [101, 308], one controlling lateral motion (own vehicle's yaw rate is used) and the other controlling longitudinal motion, to determine the speed and acceleration of detected objects. Later, they used an $\alpha - \beta$ tracker (a simplified Kalman filter with pre-estimated steady-state gains and a constant velocity model) based on the bounding box representation from their stereo verification phase [113, 114]. In this case, the Euclidean distance between bounding box centroids, shape dissimilarities (to avoid multiple tracks for single objects) and the Chamfer distance (to avoid multiple objects assigned to single tracks) are the used cues. Other researchers using Kalman filters are Bertozzi et al. [26], who use detections overlapping to merge tracks, Binelli et al. [29], who complement the predictions with egomotion computed from velocity and yaw sensors, and Grubb et al. [130], who use Bayesian probability to provide certainty, trajectory and speed of pedestrians over time as a complement to Kalman.

Particle filters are the second most extended approach. Giebel et al. [124] use them to track multiple objects in 3D (in this case the cues are silhouette, texture and stereo). Philomin et al. [250] use the Condensation method [154] (a variant of particle filters) to track silhouettes approximated by B-Splines. Following a track-before-detect paradigm, Arndt et al. [10] employ particle filters by coupling the tracking algorithm to a cascade classifier [309]. Although applied only to static scenes, it is also worth mentioning the approach in [37] given its promising results making use of a tracking-by-detection approach based on a particle filter and a texture- and color-based on-line model.

There are other approaches which do not use these two filters. Leibe et al. [180] propose the use of a color model and what they refer to as the *event cone*, i.e., the space-time volume in which the trajectory of a tracked object is sought. The authors claim that although this proposal relies on the same equations as the Kalman filter, it is superior to it in the sense that object state estimation may be based on several previous steps, and multiple trajectories for the observed data can be evaluated. Zhang et al. [338] propose the use of network flows to optimize association of detections to tracks. A min-cost flow algorithm is used to perform the detection-track association, and an explicit occlusion model is used to control long-term occlusions.

Coupled detection-tracking frameworks, i.e., modules sharing information, have been tested on detection in crowded scenarios. Gammeter et al. [108] perform multi-body tracking by combining the ISM detector [182] and the stereo-odometry based tracker of [88]. Each trajectory is passed to a single-person articulated tracker, which estimates the 3D pose and dynamics of each individual. Andriluka et al. [6] detect targets using a part-based detector and then use a Gaussian process latent variable model to compute the temporal consistency of detections over time. Singh et al. [278] use the output from the part-based detector in [328] to initialize tracklets (short track segments of high confidence detections) and residuals (low confidence detections). The tracklet descriptors (based on color, motion and 3D height) and tracklet paths (using multiple hypotheses) are then associated within a global optimization framework.

Mitzel et al. make use of a level-set tracker [212], i.e., a contour-based tracker that tracks the very dynamic pose variations of pedestrians (Fig. 4.4). The tracker is initialized by the level-set segmentation and the estimated 3D ground plane position of a new detection (using fastHOG [257]). A constant-velocity Kalman Filter is employed to estimate the 3D trajectory and use color and depth to represent each track. The same authors propose a depth-based candidates generation that is coupled with the last frame tracks, being one of the first approaches that explicitly define a feedback between tracking and candidates generation modules [214]. It is also worth mentioning that the central idea of Mitzel's works is to limit the use of the classifier only to specific frames and candidates, instead of classifying all the candidates in all the frames as in a traditional tracking-by-detection approach.

One of the most recent works, proposed by Wojek et al. [324], uses a more sophisticated tracking reasoning. First, 3D tracklet models are created for each detected

Fig. 4.4 Tracked torso shape becomes too small because of illumination changes, which triggers the re-initialization of the level-set-based shape model [212]. Image courtesy of E. Horbert (University of Aachen)

object, and greedy data association perform the short-term matching between object instances. Then, these tracklets are used in a long-term data association process that makes use of a hidden-markov-model.

4.4 Application

The application module takes high-level actions based not only on pedestrians' position, direction and velocity but also on the existence of other obstacles in the scene, the vehicle parameters (e.g., braking time, speed) and the driver attention. Fields such as automatics, artificial intelligence and human-machine interaction are combined with computer vision in this module.

This stage can be divided into two different subtasks: situation analysis and actions taking. The former one is focused on the particular driving situation by analyzing pedestrians behavior, studying the probabilities of collision, etc. It is aimed at predicting possible dangerous situations by making use of the data provided by the previous modules. The latter one makes use of the prediction to take the *front-end* action to control counter measures, for example to warn the driver or the pedestrians, decelerate the car or even taking non-reversible active actions as the deployment of external airbags.

One first approach to infer basic pedestrian behavior is to estimate the walking direction of pedestrians [277], which is a potential cue to evaluate collision risk. Tsuji et al. [297] compute the relative moving vectors of pedestrians, which together with yaw and speed information of the vehicle is used to judge possible collisions. Fuerstenberg et al. [105] use laserscanners to predict the time to collision of pedestrians entering the so-called *region of no scape*, i.e., the region where a crash is unavoidable. These techniques, along with other surveillance-oriented pedestrian behavior analysis techniques, are briefly reviewed in [109]. Recently, Keller et al. [167] go one step beyond by estimating the probability that a detected pedestrian crosses the street.

There are few publications focused on the system's active part: action taking. Here the systems describe also the situation analysis, since action taking makes use of its output. In [198] it is described how in the context of the SAVE-U project two types of actions are implemented at the application level: acoustical driver warning and automatic braking. These counter measures are applied following a three-phases strategy:

- *Phase 1: Early detection.* The system detects and tracks all pedestrians in front of the vehicle (within the sensor coverage area), but none of the protection measures are activated yet.
- *Phase 2: Acoustical driver warning.* A pedestrian is detected to enter the vehicle's trajectory, but there is no risk of an immediate collision yet. The driver is alerted by an acoustical signal about this potentially dangerous situation.

- *Phase 3: Automatic braking.* A high risk of a collision has been identified. The vehicle is decelerated in order to avert the collision or, in the case that the collision is unavoidable, mitigate the impact.

Given both the difficulty of estimating the speed of pedestrians and their unpredictable behavior (e.g., suddenly start or stop walking, even change direction) the system takes decisions only based on pedestrian position and direction but not speed.

In [297], Tsuji et al. follow a different approach. In this case, the authors make tests with two configurations for nighttime scenarios. The first one consists of a *head-up-display* (HUD) that projects the acquired FIR images in the windshield (just above the steering wheel in front of the driver) plus an additional voice assistance. The other configuration consists of a LCD screen mounted over the gearbox together with the voice assistance. Tests highlight that the first configuration is more effective, aiding the driver to initiate collision avoidance in about one second. It is worth noting that in this case the final decision is left to the driver, so no automatic actions are applied on the vehicle.

Graf et al. [128] propose two human machine interfaces for night driving. The first one consists in a small display on top of the steering wheel that just turns on when pedestrians are present in the road, fading out when there is no danger. The second does not require the driver to focus his/her attention on a display since it consists in a bar of LEDS placed over the steering wheel that turn on when pedestrians are detected.

Keller et al. [164] present an active system able to brake and perform evasive steering depending on the situation. Cues from situation analysis (pedestrian trajectory) and driver monitoring (braking, steering, etc.) are taken into account to control the vehicle. Interesting experiments performed on 22 test scenarios using both pedestrian and a dummy are performed.

It is worth highlighting a very interesting experiment performed by Källhammer et al. [161] on the factors that influence the driver's acceptance of pedestrian alerts in a PPS. This kind of experiments are crucial when embedding the aforementioned actions taking in commercial vehicles given that issuing too many alerts may annoy the driver and potentially he/she disconnects the system. The experiment analyzes the acceptance of alerts with respect to pedestrians' position and motion, location (urban, suburban, locale) and road geometry. The main conclusion is that alerts are more accepted when pedestrians are close to or moving toward the vehicle's trajectory, i.e., when they can get into the path of the vehicle. This seems to match the intuition that the most relevant factor when deciding to inform the driver is the danger of runover rather than providing accurate pedestrian location or road geometry.

4.5 Real-Time

As mentioned in Chap. 1, real-time is one of the crucial aspects in PPS. If a system with high detection rates is not capable of reaching real-time performance it will not be useful in a real vehicle. A proof of this is that even in the first PPSs the computational time aspect was already taken into account [241]. There are three ways of accelerating the system: optimizing the candidates generation, optimizing the classification and porting the bottleneck algorithms (which almost always corresponds to the features computation) to hardware implementations. A fourth way could be to combine different approaches.

Candidates generation evolved from sliding windows to more advanced methods such as road scanning or stixels with the main objective of reducing the computation time (Chap. 2). For example, road scanning can reduce the number of candidates from millions to even thousands, which represents a significant computation reduction. However, the generation process can be accelerated even by omitting the use of ADAS-specific information: e.g., by optimizing the traditional sliding windows. Lampert's *efficient subwindow search* [174] performs a branch-and-bound search that bounds the classifier output which is capable of reducing the candidates from billions to tens of thousands (the authors assume a 1-pixel window stride). Pedersoli's *discriminant multiresolution cascade* [246] performs a coarse-to-fine evaluation of the classifier (HOG-based) which makes the system 23 times faster. Dollár's [71] *fastest pedestrian detector in the west*, by using one pyramid classifier per scale octave, is able to provide the same detection performance as image pyramid-based classifiers at the same speed as when using pyramid classifier features like Haar features. Dollár et al. [69] also proposed crosstalk cascades, which take advantage of the fact that neighbor candidates have similar classifier response. This lets the system activate or inhibit the evaluation of candidates in a cascade-based manner.

The second approach, optimizing the classification, consists in accelerating the computation of the features or the classifier. One of the traditional approaches to accelerate features is by the use of the integral image. Haar-like features, edge orientation histograms or the recent integral channels [72] are examples. These features can be directly resized to adapt to multi-scale detection without the need of constructing a pyramid-scale of the input image. Another way is to accelerate specific aspects of them, such as Felzenszwalb's HOG [92], which reduces the number of dimensions by the use dimensionality reduction, the dominant orientation templates (DOT) [288], which provide a binary output, or reengineering the features as in [77], in which the authors use the Fourier transform properties to accelerate any linear filter, thus achieving speedups factors proportional to the filter's sizes. Regarding classifier optimization, in [241] the classifier speed was optimized by selecting the most representative Haar-like filters characterizing a pedestrians (only using 29 from the 1 326 original ones). The popular classifier cascades [310] only process the whole classifier (i.e., evaluates all the cascade levels) when the evaluated window corresponds to a pedestrian.

The third approach correponds to the most popular trend nowadays: to make use of parallel computation via GPUs. As an example, different implementations of the HOG descriptor have been implemented by using GPU [257, 321, 339].[1]

---

[1] Some approaches not using GPU have also been proposed. For instance, [138], in which the authors make use of SSE operations to accelerate the computation of DOT features. However, GPUs are the most promising and modern trend to optimize algorithms via hardware.

Furthermore, the particle filters, used for tracking, have also been implemented using the GPU [203]. However, while this approach is convenient for research experiments and prototypes, it is still to be clarified if the power consumption of GPUs is appropriate for an on-board vehicle computer or other approaches such as FPGAs better suit the computational speedup needs.

As a final note, it is worth to highlight Benenson's *100 fps pedestrian detector* [17, 18], which combines the three approaches to reach real-time. First, the candidates generation method uses a multi-scale model to search pedestrians in a single-scale image, which improves [71]. In addition, the authors make use of a stixels-based algorithm that avoids the explicit computation of a depth map. Second, they use cascade-like classifier, which early discards a candidate if its score does not reach a certain value. Finally, GPU is used to accelerate all the proces (both stixels and features).

4.6 Discussion

Next we discuss important aspects regarding the system modules described in this chapter and real-time.

HDR sensors provide the possibility of obtaining highly contrasted images in outdoor scenarios. This technology will be of crucial importance in PPSs in order to avoid the over/under-saturated regions that are typically seen in ADAS imaging. In fact, many of the failures of current detection algorithms correspond to poorly contrasted images (see datasets in Chap. 5), so this technology will undoubtedly benefit the system performance. Furthermore, HDR cameras cover the VS+NIR spectrum so they may also be useful for nighttime vision. An unexplored possibility for future ADAS preprocessing research is to adjust the camera parameters when the car is entering a tunnel [21, 53] and even change the system's behavior according to the scenario (estate housing, city center, suburbs, etc.) by the use of scene classification [176].

The recent development and commercialization of platforms using ADAS has brought a new set of requirements that had been seldom taken into account until now. Being stereo camera calibration a crucial aspect in any of these systems, i.e., any decalibration produced because of a hit or even of the sun light's heat can affect the stereo reconstruction, thus novel research in auto-calibration is useful. The basic requirements to be fulfilled by an auto-calibration system are the following: (1) speed, since it has to be operational as soon as the car is started; (2) auto-manteinance, as the system has to be able to detect when a recalibration is needed; and (3) robustness, as the system's stereo reconstruction has to perform as the first day. The most problematic point is the last one, as no human supervision is performed after the recalibration is carried out.

Regarding the verification process, authors often refer to the described techniques as a two-step detection process, due to the fact that verification algorithms are tied to the classification output, i.e., to the characteristics of the classification's false positives. For instance, a classifier that fails to discard trees will not gain much benefit from a verifier that just distinguishes vertical regions in 3D. It is important to note that stereo information tends to be used as long as the classification is based on 2D. In addition, it is reasonable to expect that as more cues are used in the verification process, the results will be richer, e.g., stereo imaging may be combined with classification confidence, symmetry or gait. It should also be noted that the use of verification after tracking presents an interesting approach, since common movement-based techniques (e.g., gait pattern analysis) used in surveillance may potentially be applied. The disadvantage of this approach is that this procedure is limited to walking pedestrians with clearly visible legs. This restriction usually implies that the pedestrian is close to the camera, which means that the latency of the analysis is a very important issue.

With respect to the refinement, the windows clustering topic seems to have arrived to its limit in terms of computational time and performance. Nevertheless, this limit comes from the traditional idea of seeing the clustering as an independent step from the other modules. Accordingly, a promising future research topic is its explicit combination with other modules such as candidates generation or classification, as already happens with silhouette extraction [182]. Finally, the future of silhouette extraction is linked to one of the hot topics in computer vision nowadays: image segmentation.

Tracking has the role of transforming a pedestrian detector into an almost fully operative protection system (with the application module help). However, as can be seen in this book, the number of approaches on this topic is far smaller than on other modules as classification. It can be said that in the recent years it has started to receive the deserved attention from the scientific community by presenting algorithms specifically designed for PPS and not just applying the standard filters. A clear fact is that Kalman filter is the preferred algorithm, and the cues range from simple 2D window localization to color, silhouette, texture or 3D information. Another important point is that coupled detection-tracking algorithms represent a promising way to use richer tracking cues, e.g., tracking independently detected body parts instead of complete, rigid pedestrian silhouette models, and even the potential of tracking unknown objects (i.e., not belonging to a traditional category) as trolleys or wheelchairs [213]. In any case, there is a lot of room for improvement in tracking. Generic topics such as dealing with partial/full occlusions or crowds are a hot unsolved topic of research nowadays, while other specific to ADAS, such as dealing with very dynamic scenes, the lack of color information or the aid of stereo in tracking have been rarely published. Very interesting approaches proposing the use of 2D information [324] to perform 3D scene understanding instead of 3D information [180] are in the latest trends. The inclusion of not only 3D but of other modalities (e.g., FIR, active sensors, etc.) in these newest works would probably improve the system's performance. Furthermore, there exists very little comparison between the different tracking algorithms in the PPSs context, which makes it difficult to know which are the best algorithms in the literature. A comprehensive evaluation of the different tracking algorithms in standard datasets (see Chap. 5) would be very interesting.

Finally, there are many functionalities that may be implemented at the application level: automatic braking, acoustical warnings, enhanced image visualizations (e.g., FIR image on HUD), evasive steering (with knowledge of the environment), external airbags, warning hoot for pedestrians, etc. Nevertheless, even though the application level is based on almost abstract and filtered information, the functionality implementation is far from being trivial. For instance, warning signals shall be generated early to let the driver react properly; however, since pedestrians have a very high degree of dynamic freedom they can change their moving direction in a second, making it difficult to choose the precise time to deploy the warning. Another example is the true positive vs false alarms rate. Too many false alarms could make the driver distrust the system [161]. On the contrary, drivers could unconsciously reduce their attention if they delegate too much responsibility to the system. In this sense, high level reasoning combining systems looking inside the car (e.g., driver monitoring [12]) with the forward-looking proposals (like PPSs) seems to be the option to follow. For example, when the driver is aware of what happens in the scene, the so-called *driver in the loop*, the system should not deploy any warning. It is clear that different research areas such as psychology or human-machine interaction have their relevance in this module. There are some papers addressing this topic, which is progressively getting more attraction from researchers, once that PPS have gained some maturity. Trivedi et al. [244, 296] present systems that analyze the driver's analysis and intentions. This lets them construct an application module with a wider-perspective view: it uses information from the environment, vehicle and driver. Rauch et al. [136] analyze the two main causes of pedestrian dissattention, drowsiness and distraction. The analysis is very comprehensive, detailing aspects such as direct drowssiness detection via blinking analysis, indirect detection via steering activity or lane keeping, and the different distraction causes.

Furthermore, topics such as accident analysis or safety engineering are completely linked to the role of PPSs and ADAS in general. These latter are also out of the scope of this book, so we refer to publications such as the Injury Research Journal, Journal of Accident Analysis and Prevention, the Journal of Safety Research, the International Journal of Crashworthiness, etc.

Chapter 5
Datasets and Benchmarking

In this chapter we describe the existing datasets and evaluation protocols for pedestrian detectors.

5.1 Datasets

The increasing interest in pedestrian detection during the recent years has led to the appearance of many new pedestrian datasets. Initially, the datasets were aimed at evaluating the new classifiers presented (e.g., MIT, INRIA, Daimler-Classification, CVC-01), and then they started evaluating the complete detection results (e.g., TUD, Daimler-Detection, Caltech, CVC-02-System). In Table 5.1 we summarize the existing pedestrian datasets so far, pointing out their most significant aspects. The reader may find specific technical details on the pedestrian sizes, number of examples, etc. for some of the sets in the experimental survey by Dollár et al. [74].

MIT dataset is the first pedestrian detection focused dataset, and contains a thousand of cropped pedestrian images. The secondly published set, INRIA, is still a very used set not only for PPSs context but also for generic human detection (e.g., surveillance, database retrieval, etc.). It contains high resolution photographs with standing people in different scenarios. Later, the Daimler-Classification set started the way for bigger and improved datasets. This dataset is the first ADAS-oriented public one, meaning that the images are taken from an on-board camera. Soon, researchers realized that not only the classification is important but also the whole detection system, which led to datasets containing full frames (not cropped windows). Together with this new concept, the datasets included video sequences. The Daimler-Detection and Caltech sets are the most relevant. The last trend consists in including stereo data (CVC-02 and Daimler-Stereo datasets) and occlusion labels (Caltech and Daimler-Stereo). An interesting point that should be the direction for future sets is to provide tracking data (providing a single id for pedestrian track) and vehicle information

D. Gerónimo and A. M. López, *Vision-based Pedestrian Protection Systems for Intelligent Vehicles*, SpringerBriefs in Computer Science, DOI: 10.1007/978-1-4614-7987-1_5, © The Author(s) 2014

Table 5.1 Public pedestrian datasets

	Publication		Properties							Pedestrians		
	Reference	Year	Camera	Stereo	Color	Video	Occlusion labels	Tracking labels	Vehicle data	# Train	# Test	Median height (pix)
MIT	[241]	2000	Photo		✓					924	–	128
INRIA	[60]	2005	Photo		✓					1208	566	279
USC-A	[326]	2005	Photo							–	313	98
USC-B	[326]	2005	Surveillance							–	271	90
Daimler-Classification	[220]	2006	Vehicle		✓					2400	1600	36
CVC-01	[122]	2007	Vehicle		✓					1000	–	83
USC-C	[329]	2007	Photo							–	232	108
Penn-Fudan	[313]	2007	Photo		✓					–	345	264
ETHZ	[89]	2007	Trolley	✓	✓	✓				1578	9.3k	90
NICTA	[236]	2008	Vehicle		✓					18.7k	6.9k	72
TUD	[6]	2008	Static-ground		✓	✓				400	250	218
TUD-Brussels	[325]	2009	Vehicle	✓	✓	✓				1776	1498	69
Daimler-Detection	[83]	2009	Vehicle			✓				3.9k	56k	47
Caltech	[73]	2009	Vehicle		✓	✓	✓	✓		192k	155k	48
CVC-02-Classification	[119]	2010	Vehicle		✓					1016	587	86
CVC-02-System	[119]	2010	Vehicle		✓	✓	✓			–	7.9k	78
Daimler-Multi	[81]	2010	Vehicle		✓					52k	36k	96
CVC-virtual	[200]	2010	Vehicle		✓					1678	–	96
Daimler-Stereo	[165]	2011	Vehicle	✓		✓		✓		–	56k	47

Video: video sequences, not cropped samples or isolated frames. Not counting mirrors. #Train and #Test refers to per-frame annotated pedestrians (i.e., a pedestrian is counted n times if it appears in n different frames). The number of training pedestrians only counts original pedestrian samples without mirroring (in all the sets) or jittering [83] the samples

Fig. 5.1 Test frames with their corresponding annotations of three current datasets (the frames have been cropped for publication convenience). **a** Caltech [73]. **b** CVC-02-System [119]. **c** Daimler [83]. **d** TUD-Brussels [325]

(e.g., yaw, speed, etc.), as Daimler-Stereo and Caltech do. Figure 5.1 shows some test frames with their annotations of three state-of-the-art datasets.

A special mention shall be made to CVC-Virtual dataset, which includes training examples taken from synthetic images. This idea opens the possibility of controlling the conditions of the acquisition. Furthermore, it offers unlimited groundtruth at a very low cost: once the virtual models are created, generating an automatically anno-tated video-sequence with stereo, occlusion and tracking labels is almost effortless. In this vein, some tools providing synthetic imagery containing virtual urban scenar-ios with vehicles, pedestrians and street furniture in different conditions are being developed. One example is PreScan, a commercial product presented by TASS [290] that simulates sensors data in the ADAS context.

In the area of generic human detection we also would like to highlight a few interesting datasets such as the H3D dataset [36], which includes body keypoints and region annotations of 1000 people, and the surveillance oriented *Central* ETHZ [184] and CAVIAR sets [48]. Another interesting dataset not specifically focused on pedes-trians but on ADAS is the KITTI Vision Benchmark suite [117]. This dataset is focused on aspects such as odometry, stereo reconstruction and flow computation.

5.2 Evaluation Protocols

The simplest protocol for evaluating classifier accuracy is to classify a set of samples different from the training ones, called the testing set (also referred to as groundtruth). A positive sample of the testing set (a pedestrian) is labeled as *true positive* (TP) if the classifier's confidence is higher than a threshold or *false negative* (FN) if it is lower. Similarly, a negative sample (background) is labeled as *true negative* (TN) if the classifier's confidence is lower than a threshold or *false positive* (FP) if it is higher. In order to construct the performance curve, a threshold is shifted over all the classification confidences of the testing samples (e.g., from -2 to $+2$). The typical curves for this test are receiver operating characteristic (ROC), detection-error-tradeoff (DET) and precision-recall. This protocol is called *per-window test*, and it is used in [60, 81, 85, 122, 220]. One example of per-window curve is illustrated in Fig. 5.2 (left). As can be seen, the axes of this plot are *miss rate* (MR) and *false positives per window* (FPPW). The former specifies the number of misclassified pedestrians over the total number of pedestrians in the testing set (or more strictly, the number of annotations). The latter specifies the number of incorrect detections (FP), from the set of thousands or millions of negative examples in the testing set (notice that FPPW is not a rate but an absolute value). The FPPW working range depends on the dataset, but it is usually between 10^{-1} to 10^{-6}. Notice that the axes are usually plotted in a log–log scale in order to focus on small MR and FPPW values.

Another approach is to evaluate the results on full images taking into account both the candidates generation, classification and refinement modules (when only evalu-ating classifiers the candidates generation is based on sliding windows). In this case, a detection (which is a candidate with a confidence higher than a threshold after clas-sification) is labeled as TP or FP depending on some criterion, e.g., it overlaps more than 50 % an annotated pedestrian in the groundtruth. Furthermore, if an annotation

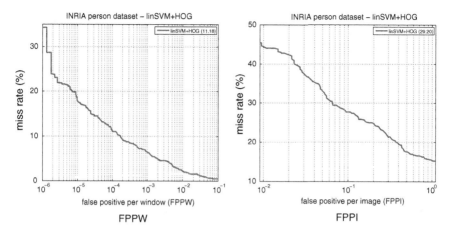

Fig. 5.2 Detection performance of a HOG+Linear SVM classifier using *per-window* (*Left*) and *per-image* (*Right*) plots

is not overlapped by any detection it is labeled as FN. This kind of protocol is *per-image test* and it is used in [59, 74, 83] and in the PASCAL VOC Challenge.[1] The working range tends to be between 10^{-2} and 10^0 false positives per image (FPPI). Figure 5.2 (right) illustrates an per-image example curve. We have mentioned the criterion of labeling a detection depending on the overlap with an annotated pedestrian. This is called PASCAL criterion. There are other criteria that are gaining popularity in PPSs. Daimler criterion [83] divides the samples in obligatory and mandatory depending on the distance and occlusion, and then use the 25 % overlap. Caltech criterion [74] extends the specification of the criterion by adding *ignore* annotation regions and crowd regions (which do not count for the performance), and detections filtering depending on size ranges. These additional constraints to compute TP/FP/etc. are an approach to make the fairest classifier evaluations as possible: e.g., to take into account detections out of range (e.g., detection at 26 m when the working range is from 0 to 25 m) matching annotations inside the range (0–25 m), which in a traditional PACAL or Daimler criterion would not be considered.

An important fact is that a high *per-window* accuracy has traditionally assumed to lead to a high *per-image* accuracy. However, this correlation has been demonstrated weak in several publications mainly because the windows tested in *per-image* are not the same as the ones tested in *per-window*. Tran et al. demonstrate this effect in [295]. In this paper, the authors present a classifier that is robust to poorly-centered pedestrian windows, which are not included in *per-window* tests but they are usual in *per-image* tests. Another significative cause of this weak correlation are parts of pedestrians (legs, arms, etc.) or too big detections framing pedestrians, which usually make the classifier peak and are normally excluded in *per-window* tests, as Dollár et al. [74] show. In this paper, the authors also point out the effect of refinement.

---

[1] http://pascallin.ecs.soton.ac.uk/challenges/VOC

In general, the best option to evaluate not only systems but also new classifiers is the *per-image test*. However, in the case of a big and well represented negative set, *per-window test* can also provide representative results similar to *per-image* [84].

Other performance measures consist of segmentation side accuracy and segmentation side efficiency, which are used to evaluate detection in the FIR by using a full-image based approach [91]; and trajectory/alarms, which take into account the tracking and application modules [114].

5.3 Discussion

In the last five years a significant evolution in pedestrian datasets has been carried out. Some traditional weak points such as the low quantity of examples in the pioneer datasets have been fixed in the modern ones. As an example, Daimler or Caltech sets contain tens of thousands of individual pedestrian annotations. The representativity of pedestrians, another traditional weak point, has also been solved in part thanks to the increase in the number of examples. However, there are still pedestrian classes such as children that are under-represented if we take into account the potential danger of run over. In this sense, some publications such as [14] claim to have used a specific children database. In summary, there is still a lot of room for improvement in the datasets in the following aspects:

- *Variability*. In addition to clothes, pose and illumination, it would be interesting to increase the variability in terms of height, distance, degree of occlusion and complexity of the background. Here virtual pedestrian synthesis [87, 200] could be very helpful. The first datasets including occlusion information have already been published (Caltech and Daimler-Stereo), which makes possible to take into account the detection or misdetection of pedestrians that were normally omitted from the evaluation.
- *Resolution*. Some databases [60] contain well-focused images of pedestrians from a photographic camera, which usually corresponds to targets close to the vehicle in a PPS. Hence, it is difficult to determine both the working distance of a given classifier and if the results can be extrapolated to detect farther targets, which tend to be blurred and contained within a small number of pixels.
- *Sensors*. Until now we have referred to visible spectrum datasets, but at the moment of writing this book no public FIR ADAS-oriented database have been presented. To the best of our knowledge, the OTCBVS[2] [64] is the only publicly available FIR database, but it is not suited to ADAS benchmarking. The same issue presents itself with active sensors such as laserscanner or radar.
- *Later-stages data*. Tracking annotation and information from the application module are surely the next data to be provided. As can be seen in the table, the only datasets that provide tracking information are the ones from Daimler and

---

[2] http://www.cse.ohio-state.edu/otcbvs-bench

Caltech. However, a proper evaluation of this stage is of big importance to develop application-level solutions. Furthermore, public benchmarkings including annotations from high-level applications such as possible vehicle actions (braking, activating external airbag, etc.) or on the potential evasive paths to be taken are needed. This kind of information, which in some cases is subject to specific vehicle parameters (braking power, steering capabilities) and external conditions (weather, visibility, etc.), is likely to be included in the datasets in the next years.

Regarding the evaluation protocols, although there exists a *de facto* standard for object detection in images, the so-called PASCAL criterion, and an increasingly used Caltech criterion for pedestrian detection, there are still some points to be improved or clarified from it. For example, whether the annotation windows' aspect ratio must be fixed (e.g., CVC-02) or not (e.g., Caltech, Daimler), which at this moment depends on the dataset, the way the tracking must be evaluated (although some approaches such as [20] are establishing as standard), or the statistical tests applied (e.g., results can improve or worsen up from one to even two or three points in detection rate as a results of the random selection of false positives during the bootstrapping, depending on the classifier).

Furthermore, although big efforts have been made to comprehensively compare algorithms [74], the studied classifiers are trained mostly using the same non-ADAS oriented dataset (INRIA) to be tested on the different sets. A common benchmark implementing the state-of-the-art algorithms, capable of performing not only testing but also training on different datasets, would probably rise new insights and would lead to new interesting proposals (e.g., algorithms combination). This will not be an easy task given that researchers normally do not make their algorithms publicly available, and in most of the cases the implementation details are not specified in the papers.

Chapter 6
Conclusions

The task of detecting, tracking and even inferring the intention of pedestrians in the scene is a very challenging topic. As a proof of the challenge it represents a plethora of research papers that have been presented during the last decades, which have progressively enriched the state of the art with faster and improved detectors. At this moment the first vehicles incorporating pedestrian detectors are being presented for mass production. In this chapter we coarsely summarize the evolution of this research field describing the most advanced current techniques and where the research community is now. Next, we detail the challenges that still need to be addressed to improve current PPSs capabilities.

6.1 State of the Research

According to Dollár et al. [74], the state-of-the-art performance is around 10 % of MR with 1 FPPI with generic human datasets [60], and in the range of 30–60 % of MR (miss-rate) with 1 FPPI (false positives per image) for ADAS-oriented sets (Daimler, Caltech, etc.). These numbers refer to systems making use of a sliding windows candidates generation approach. The attachment of more sophisticated candidates generation algorithms could improve the MR from 2 to 10 points in an ADAS-oriented set (CVC-02-Classification) [119], depending on the detection range. Regarding computation time, a typical detector using sliding windows as candidates generation algorithm takes around 1 fps in a typical computer. The current fastest proposals are Dollár's FPDW [71] (6 fps) and crosstalk [69] (35–65 fps), and Benenson's 100 fps pedestrian detector [17] (100 fps) and stixels-based [18] (150 fps), incorporating many algorithmic (features computation tricks) and hardware (e.g., fastHOG [257]) improvements. When incorporating a tracking stage, Enzweiler's et al. [86] output 2 FP trajectories every 1000 frames (feeding a 39 % MR after the refinement module) again in an ADAS dataset (Daimler-Stereo).

D. Gerónimo and A. M. López, *Vision-based Pedestrian Protection Systems for Intelligent Vehicles*, SpringerBriefs in Computer Science, DOI: 10.1007/978-1-4614-7987-1_6, © The Author(s) 2014

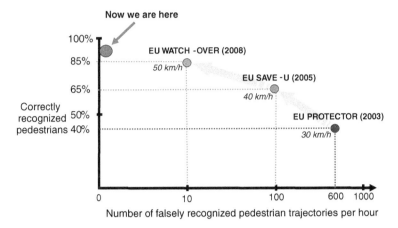

Fig. 6.1 Evolution of the Daimler stereo-vision pedestrian system performance over different EU projects in the past decade, up to 2010. Note that the number of false alarms per hour at the system level will be much lower, given that only a small subset of the considered false trajectories will lie on collision-course, and furthermore, considering additional sensors (i.e. radar) for verification. Image courtesy of Dariu Gavrila (Daimler AG)

While writing this book, Daimler announced the introduction of a stereo-based pedestrian protection system (PPS) in the Mercedes-Benz 2013 E-Class and S-Class models. The overall safety system includes fully automatic emergency braking and it works both day and night. It is the first time a major European vehicle manufacturer introduces a stereo camera in its vehicle, with benefits for the pedestrian application in terms of higher recognition rate and localization accuracy. Evaluation of accident data by Mercedes-Benz indicates that this new technology could avoid 6 % of accidents involving pedestrians and reduce the severity of a further 41 %.

Commercial PPSs, either mono-based (e.g. Volvo S60 and V60 sedans, supplied by Mobileye[1]) or stereo-based (e.g. Daimler's Mercedes-Benz E- and S-Class models), draw a more optimistic panorama, where proprietary systems' performance seems higher than suggested by the state-of-the-art performance of public research. According to Dariu Gavrila[2], who led the long-standing pedestrian detection research effort at Daimler (see Fig. 6.1), this "performance gap" can be in part explained in algorithmic terms, by a better fusion of multiple visual cues (e.g. edge vs. region features, intensity vs. depth information). Equally important is the use of much larger training datasets, enhancing pattern classification performance. Gavrila further points out that in commercial PPS that include automatic braking, there is a second (and independent) sensor component (i.e. radar), which validates the output of the video component, thus significantly reducing the number of false positives at the system level. What is more, only objects on collision course are really application relevant. An illustrative overview of the Daimler pedestrian research (incl. video footage) is

---

[1] http://www.mobileye.com

[2] http://www.gavrila.net

given in his "Smart Cars for Safe Pedestrians" keynote at the Intelligent Vehicles Symposium 2012, available from YouTube [259].

6.2 Future Challenges

The vast majority of works presented in this book are focused on detecting standard-upstanding-relatively-isolated pedestrians[3] in urban scenes at day-time with sunny or cloudy weather from a camera placed in the windshield or the roof of a standard sedan. There are some works tackling nighttime detection using NIR/FIR sensors. It is clear that robustly solving the task for the mentioned circumstances and context is of mandatory importance and would indeed represent a huge step in the field. But this does not have to hide the fact that there is still room for exploration. Next we list some detection cases that have been barely or not researched at all.

Children are much smaller than standard pedestrians, which can make the system fail in the case that no appropriate candidate window was generated for him/her. Luckily, the appearance of an upstanding child is similar to an adult at farther distance, i.e., the aspect ratio tends to be similar and only minor physiological characteristics makes them different, so children are sometimes detected too. However, it is clear that the pose of children is indeed much more changing, i.e., they are more likely to be running or laying on the ground. Up to our knowledge, the only papers specifically addressing the case of children are [14, 218], in which the authors even make use of a specific dataset to test children detection.

A relevant research line in PPSs (Pedestrian Protection Systems) is partial occlusion handling. In urban scenarios, partial occlusions occur very often (e.g., pedestrians can suddenly appear from behind a parked vehicle, there can be self-occlusions from umbrellas, bags, etc.)[4] but at the same time they are a typical source of danger. Up to now, it is usually assumed that the pedestrians' full body is visible even in part-based methods given that they need a reliable classification of parts. Some articles have proposed techniques to handle occlusions, but more research on standard datasets with annotated occlusions (categorized by type, e.g., by a car, by other pedestrians; and by danger level) and even on the evaluation of occluded targets is needed. Furthermore, the benefits of exploiting multiple cues such as optical flow, appearance or stereo in the detection of potential occluded pedestrians need to be researched further.

The environmental conditions (weather and outdoor illumination) are one of the common problems for all ADAS. For instance, at the moment we are not aware of publicly available datasets for pedestrian detection in the presence of rain, snow. Fortunately, there are interesting proposals such as raindrop detection [134], rain or snow detection (in static sequences) [34] which can help to adapt pedestrian detection to such adverse driving conditions.

---

[3] 1.75 m high with average complexion and western-like clothes.

[4] Dollár et al. [74] state that 70 % of pedestrians are occluded in at least one frame of their trajectory.

Although the research has been mainly focused on daytime, there is also significant work in nighttime based on other modality: FIR. At this moment there exist series production vehicles with displays showing FIR images of the scene, so we can expect the companies to take advantage of these expensive sensors for PPSs. However, although FIR based research is interesting, NIR sensors perhaps deserve more attention, if we take into account some interesting aspects. First, modern headlight systems cover the visible near infrared ranges and have motion capabilities, i.e., the lamps move according to the vehicle direction and even are able to spot a pedestrian. Second, NIR images are quite similar to VS images, so daytime detectors could be easily adapted to these new images than to FIR imagery. Third, there are other ADAS applications as lane departure warning or traffic sign recognition for which FIR imagery does not seem appropriate, in contrast to NIR.

Since the aim of a PPS is to assist the drivers, not to substitute them, it is not necessary to bother the driver with information if he/she is already paying attention to the road. On the contrary, the PPS can warn the driver about risky pedestrians on road areas that he/she are not monitoring (e.g., pedestrians suddenly appearing from a lateral direction). This process corresponds to the concept of *driver in the loop*, in which the current state of the driver (e.g., driver is aware, driver is distracted, etc.) is taken into account. In spite of some excellent approaches in this lines (Chap. 4), more research into driver monitoring (e.g., developing databases of synchronized driver and outdoor images) and psychological aspects (i.e., the danger of drivers intentionally paying less attention because of the PPS) is required.

Moreover, the understanding of the immediate scene ahead the vehicle as well as performing high level risk analysis will bring us better PPSs. For instance, by predicting the trajectory and intention of pedestrians [167] and knowing the free road surface in front of the vehicle, would allow going from braking maneuvers to even move sophisticated evasive ones [164].

Although the book is focused on PPSs, it is important to remind that they will be part (sub-systems) of a complete ADAS containing vehicle detectors, traffic sign recognizers, lane departure warning, etc. This has two main outcomes. First, it is worth noticing that several algorithms could be shared by different sub-systems, such as the ground plane estimation or the obstacle detection, which could be used by candidates generation modules in pedestrian and vehicle detection. Furthermore, the cooperation of different sub-systems, for example by the use of multi-class classification paradigms, can increase the robustness. Second, the hardware selection (cameras, computation system, etc.) will be decided by taking into account the requirements of several sub-systems and not only the pedestrian detection needs. This can restrict the cameras' resolution and FOV, having color or monochrome imagery available, etc. For example, in some applications grayscale images are preferred because Bayer-pattern based color cameras tend to have lower sensibility (Chap. 4), but color is still a very relevant cue for target tracking. In summary, it may be the case that the PPS sub-system cannot assume the best image acquisition conditions for its task.

References

1. Abramson, Y., Freund, Y.: SEmi-automatic VIsuaL LEarning (SEVILLE): a tutorial on active learning for visual object recognition. In: IEEE Conf. on Computer Vision and Pattern Recognition. San Diego, CA, USA (2005)
2. Agarwal, S., Awan, A., Roth, D.: Learning to detect objects in images via a sparse, part-based representation. IEEE Trans. on Pattern Analysis and, Machine Intelligence **26**(11), 1475–1490 (2004)
3. Álvarez, J., Gevers, T., LeCun, Y., López, A.: Road scene segmentation from a single image. In: European Conf. on Computer Vision. Firenze, Italy (2012)
4. Álvarez, S., Sotelo, M., Parra, I., Llorca, D., Gavilán, M.: Vehicle and pedestrian detection in esafety applications. In: World Congress on Engineering and Computer Science. San Francisco, CA, USA (2009)
5. Andriluka, M., Roth, S., Schiele, B.: Monocular 3d pose estimation and tracking by detection. In: IEEE Conf. on Computer Vision and Pattern Recognition. San Francisco, CA, USA (2008)
6. Andriluka, M., Roth, S., Schiele, B.: People-tracking-by-detection and people-detection-by-tracking. In: IEEE Conf. on Computer Vision and Pattern Recognition. Anchorage, AK, USA (2008)
7. Andriluka, M., Roth, S., Schiele, B.: Pictorial structures revisited: people detection and articulated pose estimation. In: IEEE Conf. on Computer Vision and Pattern Recognition. Miami, FL, USA (2009)
8. Anwer, R., Vázquez, D., López, A.: Color contribution to part-based person detection in different types of scenario. In: Int. Conf. on Computer Analysis of Images and Patterns. Seville, Spain (2011)
9. Anwer, R., Vázquez, D., López, A.: Opponent colors for human detection. In: IAPR Iberian Conf. on Pattern Recognition and Image Analysis. Las Palmas de Gran Canaria, Spain (2011)
10. Arndt, R., Schweiger, R., Ritter, W., Paulus, D., Lohlein, O.: Detection and tracking of multiple pedestrians in automotive applications. In: IEEE Intelligent Vehicles Symposium. Istanbul, Turkey (2007)
11. Ashton, S., Mackay, G.: Benefits from changes in vehicle exterior design. Proceedings of the Society of Automotive Engineers pp. 255–264 (1983)
12. http://www.awake-eu.org
13. Badino, H., Franke, U., Pfeiffer, D.: The stixel world-a compact medium level representation of the 3d-world. In: DAGM Symposium. Jena, Germany (2009)
14. Bar-Hillel, A., Levi, D., Krupka, E., Goldberg, C.: Part-based feature synthesis for human detection. In: European Conf. on Computer Vision. Crete, Greece (2010)

D. Gerónimo and A. M. López, *Vision-based Pedestrian Protection Systems for Intelligent Vehicles*, SpringerBriefs in Computer Science, DOI: 10.1007/978-1-4614-7987-1, © The Author(s) 2014

15. Barrera, F., Lumbreras, F., Sappa, A.: Multimodal stereo vision system: 3d data extraction and algorithm evaluation. IEEE Journal of Selected Topics in Signal Processing (in press) (2012)
16. Belongie, S., Malik, J., Puzicha, J.: Shape matching and object recognition using shape contexts. IEEE Trans. on Pattern Analysis and, Machine Intelligence **24**(4), 509–522 (2002)
17. Benenson, R., Mathias, M., Timofte, R., Van Gool, L.: Fast stixels estimation for fast pedestrian detection. In: European Conf. on Computer Vision—Workshop on Computer Vision in Vehicle Technology: From Earth to Mars at ECCV. Firenze, Italy (2012)
18. Benenson, R., Mathias, M., Timofte, R., Van Gool, L.: Pedestrian detection at 100 frames per second. In: IEEE Conf. on Computer Vision and Pattern Recognition. Providence, RI, USA (2012)
19. Berg, T., Sorokin, A., Wang, G., Forsyth, D., Hoeiem, D., Endres, I., Farhadi, A.: It's all about the data. Proceedings of the IEEE **98**(8), 1434–1452 (2010)
20. Bernardin, K., Stiefelhagen, R.: Evaluating multiple object tracking performance: the clear mot metrics. EURASIP Journal on Image and Video Processing (2010)
21. Bertozzi, M., Broggi, A., Boccalini, G., Mazzei, L.: Fast vision-based road tunnel detection. In: IEEE Int. Conf. on Image Processing. Singapore (2004)
22. Bertozzi, M., Broggi, A., Carletti, M., Fascioli, A., Graf, T., Grisleri, P., Meinecke, M.: IR pedestrian detection for advanced driver assistance systems. In: DAGM Symposium. Magdeburg, Germany (2003)
23. Bertozzi, M., Broggi, A., Chapuis, R., Chausse, F., Fascioli, A., Tibaldi, A.: Shape-based pedestrian detection and localization. In: IEEE Int. Conf. on Intelligent Transportation Systems. Shangai, China (2003)
24. Bertozzi, M., Broggi, A., Del Rose, M., Felisa, M.: A symmetry-based validator and refinement system for pedestrian detection in far infrared images. In: IEEE Int. Conf. on Intelligent Transportation Systems. Seattle, WA, USA (2007)
25. Bertozzi, M., Broggi, A., Fascioli, A., Graf, T., Meinecke, M.: Pedestrian detection for driver assistance using multiresolution infrared vision. IEEE Trans. on Vehicular Technology **53**(6), 1666–1678 (2004)
26. Bertozzi, M., Broggi, A., Fascioli, A., Tibaldi, A., Chapuis, R., Chausse, F.: Pedestrian localization and tracking system with Kalman filtering. In: IEEE Intelligent Vehicles Symposium. Parma, Italy (2004)
27. Bertozzi, M., Broggi, A., Graf, T., Meinecke, M.: Pedestrian detection in infrared images. In: IEEE Intelligent Vehicles Symposium. Columbus, OH, USA (2003)
28. Bertozzi, M., Broggi, A., Lasagni, A., Del Rose, M.: Infrared stereo vision-based pedestrian detection. In: IEEE Intelligent Vehicles Symposium. Las Vegas, NV, USA (2005)
29. Binelli, E., Broggi, A., Fascioli, A., Ghidoni, S., Grisleri, P., Graf, T., Meinecke, M.: A modular tracking system for far infrared pedestrian recognition. In: IEEE Intelligent Vehicles Symposium. Las Vegas, NV, USA (2005)
30. Bishop, C.: Neural networks for pattern recognition. Oxford University Press (1995)
31. Bishop, R.: Intelligent Vehicle Technologies and Trends. Artech House, Inc. (2005)
32. Bombini, L., Cerri, P., Grisleri, P., Scaffardi, S., Zani, P.: An evaluation of monocular image stabilization algorithms for automotive applications. In: IEEE Int. Conf. on Intelligent Transportation Systems. Toronto, Canada (2006)
33. Borgefors, G.: Distance transformations in digital images. Computer Vision, Graphics and Image Processing **34**(3), 344–371 (1986)
34. Bossu, J., Hautière, N., Tarel, J.: Rain or snow detection in image sequences through use of a histogram of orientation of streaks. Int. Journal on Computer Vision **93**(3) (2011)
35. Bourdev, L., Maji, S., Brox, T., Malik, J.: Detecting people using mutually consistent poselet activations. In: European Conf. on Computer Vision. Crete, Greece (2010)
36. Bourdev, L., Malik, J.: Poselets: Body part detectors trained using 3d human pose annotations. In: Int. Conf. on Computer Vision. Kyoto, Japan (2009)
37. Breitenstein, M., Reichlin, F., Leibe, B., Koller-Meier, E., Van Gool, L.: Online multiperson tracking-by-detection from a single, uncalibrated camera. IEEE Trans. on Pattern Analysis and, Machine Intelligence **33**(9), 1820–1833 (2011)

38. Broggi, A., Bertozzi, M., Fascioli, A.: Self-calibration of a stereo vision system for automotive applications. In: IEEE Int. Conf. on Robotics and Automation. Seoul, Korea (2001)
39. Broggi, A., Bertozzi, M., Fascioli, A., Sechi, M.: Shape-based pedestrian detection. In: IEEE Intelligent Vehicles Symposium. Dearborn, MI, USA (2000)
40. Broggi, A., Fascioli, A., Carletti, M., Graf, T., Meinecke, M.: A multi-resolution approach for infrared vision-based pedestrian detection. In: IEEE Intelligent Vehicles Symposium. Parma, Italy (2004)
41. Broggi, A., Fascioli, A., Fedriga, I., Tibaldi, A., Rose, M.D.: Stereo-based preprocessing for human shape localization in unstructured environments. In: IEEE Intelligent Vehicles Symposium. Columbus, OH, USA (2003)
42. Broggi, A., Fascioli, A., Grisleri, P., Graf, T., Meinecke, M.: Model-based validation approaches and matching techniques for automotive vision based pedestrian detection. In: IEEE Conf. on Computer Vision and Pattern Recognition—Workshop on Object Tracking and Classification in and Beyond the Visible Spectrum. San Diego, CA, USA (2005)
43. Broggi, A., Fedriga, R., Tagliati, A., Graf, T., Meinecke, M.: Pedestrian detection on a moving vehicle: an investigation about near infra-red images. In: IEEE Intelligent Vehicles Symposium. Tokyo, Japan (2006)
44. Broggi, A., Grisleri, P., Graf, T., M.M. Meinecke: A software video stabilization system for automotive oriented applications. In: Vehicular Technology Conf. Dallas, TX, USA (2005)
45. C. Zach, M.N., Frahm, J.: Continuous maximal flows and wulff shapes: Application to mrfs. In: IEEE Conf. on Computer Vision and Pattern Recognition. Miami, FL, USA (2009)
46. Caballero, H.: Pedestrian detection refinement via non-explicit shape models. Tech. rep., Computer Vision Center (2009)
47. Cao, H., Yamaguchi, K., Naito, T., Ninomiya, Y.: Pedestrian recognition using second-order HOG feature. In: Asian Conf. on Computer Vision. Xi'an, China (2009)
48. http://homepages.inf.ed.ac.uk/rbf/CAVIARDATA1
49. Chambolle, A., Pock, T.: A first-order primal-dual algorithm for convex problems with applications to imaging. Journal of Mathematical Imaging and Vision **40**(1), 120–145 (2011).
50. Chan, C., Bu, F., Shladover, S.: Experimental vehicle platform for pedestrian detection. Tech. rep., Institute of Transportation Studies, University of California (2006)
51. Chen, Y., Chen, C.: A cascade of feed-forward classifiers for fast pedestrian detection. In: Asian Conf. on Computer Vision. Tokyo, Japan (2007)
52. Chen, Y., Chen, C.: Fast human detection using a novel boosted cascading structure with meta-satges. IEEE Trans. on Image Processing **17**(8), 1452–1464 (2008)
53. Claus, C., Shin, H., Stechele, W.: Tunnel entrance recognition for video-based driver assistance systems. In: Semantic Multimodal analysis of digital media. Budapest, Hungary (2006)
54. Cohn, D., Atlas, L., Ladner, R.: Improving generalization with active learning. Machine Learning **15**(2), 201–221 (1994)
55. Comaniciu, D.: An algorithm for data-driven bandwidth selection. IEEE Trans. on Pattern Analysis and, Machine Intelligence **25**(2), 281–288 (2003)
56. Crow, L.: Summed area tables for texture mapping. In: ACM SIGGRAPH. Minneapolis, MN, USA (1984)
57. Cui, X., Liu, Y., Shan, S., Chen, X., Gao, W.: 3d haar-like features for pedestrian detection. In: IEEE Int. Conf. on Multimedia & Expo. Bejing, China (2007)
58. Dai, S., Yang, M., Wu, Y., Kasatggelos, A.: Detector ensemble. In: IEEE Conf. on Computer Vision and Pattern Recognition. Minneapolis, MN, USA (2007)
59. Dalal, N.: Finding people in images and videos. PhD Thesis, Institut National Polytechnique de Grenoble/ INRIA Rhône-Alpes (2006)
60. Dalal, N., Triggs, B.: Histograms of oriented gradients for human detection. In: IEEE Conf. on Computer Vision and Pattern Recognition. San Diego, CA, USA (2005)
61. Dalal, N., Triggs, B., Schmid, C.: Human detection using oriented histograms of flow and appearance. In: European Conf. on Computer Vision. Graz, Austria (2006)
62. Dang, T., Hoffmann, C.: Stereo calibration in vehicles. In: IEEE Intelligent Vehicles Symposium. Parma, Italy (2004)

63. Dang, T., Hoffmann, C., Stiller, C.: Continuous stereo self-calibration by camera parameter tracking **18**(7), 1536–1550 (2009)
64. David, J., Keck, M.: A two-stage approach to person detection in thermal imagery. In: Workshop on Applications of Computer Vision. Breckenridge, CO, USA (2005)
65. Davis, L., Johns, S., Aggarwal, J.: Texture analysis using generalized co-occurrence matrices. IEEE Trans. on Pattern Analysis and, Machine Intelligence **1**(3), 251–259 (1979)
66. Davis, S., Diegel, S., Boundy, R.: Transportation Energy Data Book: Edition 30. Office of Energy Efficiency and Renewable Energy, U.S. Department of Energy, USA (2011)
67. Dickmanns, E., Zapp, A.: A curvature-based scheme for improving road vehicle guidance by computer vision. In: SPIE Conference on Mobile Robots (1986)
68. Divvala, S., Efros, A., Hebert, M.: How important are 'deformable parts' in the deformable parts model? In: European Conf. on Computer Vision—Workshop on Parts and Attributes. Firenze, Italy (2012)
69. Dollár, P., Appel, R., Kienzle, W.: Crosstalk cascades for frame-rate pedestrian detection. In: European Conf. on Computer Vision. Firenze, Italy (2012)
70. Dollár, P., Babenko, B., Belongie, S., Perona, P., Tu, Z.: Multiple component learning for object detection. In: European Conf. on Computer Vision. Marseille, France (2008)
71. Dollár, P., Belongie, S., Perona, P.: The fastest pedestrian detector in the west. In: British Machine Vision Conference (2010)
72. Dollár, P., Tu, Z., Perona, P., Belongie, S.: Integral channel features. In: British Machine Vision Conference. London, UK (2009)
73. Dollár, P., Wojek, C., Schiele, B., Perona, P.: Pedestrian detection: A benchmark. In: IEEE Conf. on Computer Vision and Pattern Recognition. Miami, FL, USA (2009)
74. Dollár, P., Wojek, C., Schiele, B., Perona, P.: Pedestrian detection: An evaluation of the state of the art. IEEE Trans. on Pattern Analysis and, Machine Intelligence **34**(4), 743–761 (2012)
75. Duan, G., Ai, H., Lao, S.: A structural filter approach to human detection. In: European Conf. on Computer Vision. Crete, Greece (2010)
76. Duan, G., Huang, C., Ai, H., Lao, S.: Boosting associated pairing comparison features for pedestrian detection. In: Int. Conf. on Computer Vision—Workshop on Visual Surveillance. Kyoto, Japan (2009)
77. Dubout, C., Fleuret, F.: Exact acceleration of linear object detectors. In: European Conf. on Computer Vision. Firenze, Italy (2012)
78. Economic Comision for Europe: Statistics of Road Traffic Accidents in Europe and North America, vol. LI. United Nations (2007)
79. Elzein, H., Lakshmanan, S., Watta, P.: A motion and shape-based pedestrian detection algorithm. In: IEEE Intelligent Vehicles Symposium. Columbus, OH, USA (2003)
80. Endres, I., Farhadi, A., Forsyth, D.H.D.: The benefits and challenges of collecting richer annotations. In: IEEE Conf. on Computer Vision and Pattern Recognition—Workshop on Advancing Computer Vision with Humans in the Loop. San Francisco, CA, USA (2010)
81. Enzweiler, M., Eigenstetter, A., Schiele, B., Gavrila, D.: Multi-cue pedestrian classification with partial occlusion handling. In: IEEE Conf. on Computer Vision and Pattern Recognition. San Francisco, CA, USA (2010)
82. Enzweiler, M., Gavrila, D.: A mixed generative-discriminative framework for pedestrian classification. In: IEEE Conf. on Computer Vision and Pattern Recognition. Anchorage, AK, USA (2008)
83. Enzweiler, M., Gavrila, D.: Monocular pedestrian detection: Survey and experiments. IEEE Trans. on Pattern Analysis and, Machine Intelligence **31**(12), 2179–2195 (2009)
84. Enzweiler, M., Gavrila, D.: Integrated pedestrian classification and orientation estimation. In: IEEE Conf. on Computer Vision and Pattern Recognition. San Francisco, CA, USA (2010)
85. Enzweiler, M., Gavrila, D.: A multi-level mixture-of-experts framework for pedestrian classification. IEEE Trans. on Image Processing **20**(10), 2967–2979 (2011)
86. Enzweiler, M., Hummel, M., Pfeiffer, D., Franke, U.: Efficient stixel-based object recognition. In: IEEE Intelligent Vehicles Symposium. Alcalá de Henares, Spain (2012)

87. Enzweiler, M., Kanter, P., Gavrila, D.: Monocular pedestrian recognition using motion parallax. In: IEEE Intelligent Vehicles Symposium. Eindhoven, The Netherlands (2008)
88. Ess, A., Leibe, B., Schindler, K., Van Gool, L.: A mobile vision system for robust multi-person tracking. In: IEEE Conf. on Computer Vision and Pattern Recognition. Anchorage, AK, USA (2008)
89. Ess, A., Leibe, B., Van Gool, L.: Depth and appearance for mobile scene analysis. In: Int. Conf. on Computer Vision. Rio de Janeiro, Brazil (2007)
90. Fallon, I., O'Neill, D.: The world's first automobile fatality. Accident Analysis and Prevention **37** (2005)
91. Fang, Y., Yamada, K., Ninomiya, Y., Horn, B., Masaki, I.: A shape-independent method for pedestrian detection with far-infrared images. IEEE Trans. on Vehicular Technology **53**(6), 1679–1697 (2004).
92. Felzenszwalb, P., Girshick, R., McAllester, D., Ramanan, D.: Object detection with discriminatively trained part based models. IEEE Trans. on Pattern Analysis and, Machine Intelligence **32**(9), 1627–1645 (2010).
93. Felzenszwalb, P., McAllester, D., Ramanan, D.: A discriminatively trained, multiscale, deformable part model. In: IEEE Conf. on Computer Vision and Pattern Recognition. Anchorage, AK, USA (2008).
94. Fernández, D., Parra, I., Sotelo, M., Bergasa, L., Revenga, P., Nuevo, J., Ocaa, M.: Pedestrian recognition for intelligent transportation systems. In: Int. Conf. on Informatics in Control, Automation and Robotics. Barcelona, Spain (2005)
95. Fischler, M., Elschlager, R.: The representation and matching of pictorial structures. IEEE Trans. on Computers **100**(1), 67–92 (1973)
96. http://www.fnir.eu
97. Forsyth, D., Arikan, O., Ikemoto, L., O'Brien, J., Ramanan, D.: Computational Studies of Human Motion: Part 1, Tracking and Motion Synthesis. Now publishers (2005)
98. Forsyth, D., Fleck, M.: Body plans. In: IEEE Conf. on Computer Vision and Pattern Recognition. Puerto Rico, USA (2012)
99. Franke, U., Gavrila, D., Görzig, S., Lindner, F., Paetzold, F., Wöhler, C.: Autonomous driving goes downtown. IEEE Intelligent Systems **13**(6), 40–48 (1999)
100. Franke, U., Heinrich, S.: Fast obstacle detection for urban traffic situations. IEEE Trans. on Intelligent Transportation Systems **3**(3), 173–181 (2002).
101. Franke, U., Joos, A.: Real-time stereo vision for urban traffic scene understanding. In: IEEE Intelligent Vehicles Symposium. Dearborn, MI, USA (2000)
102. Franke, U., Kutzbach, I.: Fast stereo based object detection for Stop & Go traffic. In: IEEE Intelligent Vehicles Symposium. Tokyo, Japan (1996)
103. Franke, U., Rabe, C., Badino, H., Gehrig, S.: 6D-Vision: Fusion of stereo and motion for robust environment perception. In: DAGM Symposium. Vienna, Austria (2005)
104. Freund, Y., Schapire, R.: A decision-theoretic generalization of on-line learning and an application to boosting. Journal of Computer and System Sciences **55**(1), 119–139 (1997)
105. Fuerstenberg, K.: Pedestrian protection using laserscanners. In: IEEE Int. Conf. on Intelligent Transportation Systems. Vienna, Austria (2005)
106. Fukushima, K.: Neocognitron: a neural network model for a mechanism of pattern recognition unaffected by shift in position. Biological Cybernetics **36**(4), 193–202, year = 1980
107. Fukushima, K., Miyake, S., Ito, T.: Neocognitron: a self-organizing neural network model for a mechanism of visual pattern recognition. IEEE Trans. on Systems, Man, and, Cybernetics **13**(3), 826–834 (1983)
108. Gammeter, S., Ess, A., Jäggli, T., Schindler, K., Leibe, B., Van Gool, L.: Articulated multibody tracking under egomotion. In: European Conf. on Computer Vision. Marseille, France (2008)
109. Gandhi, T., Trivedi, M.: Pedestrian protection systems: Issues, survey, and challenges. IEEE Trans. on Intelligent Transportation Systems **8**(3), 413–430 (2007)
110. Gavrila, D.: The Visual Analysis of Human Movement: a survey. Computer Vision and Image Understanding **73**(1), 82–98 (1999)

111. Gavrila, D.: Pedestrian detection from a moving vehicle. In: European Conf. on Computer Vision. Dublin, Ireland (2000)
112. Gavrila, D.: A Bayesian, exemplar-based approach to hierarchical shape matching. IEEE Trans. on Pattern Analysis and, Machine Intelligence **29**(8), 1408–1421 (2007)
113. Gavrila, D., Giebel, J., Munder, S.: Vision-based pedestrian detection: The PROTECTOR system. In: IEEE Intelligent Vehicles Symposium. Parma, Italy (2004)
114. Gavrila, D., Munder, S.: Multi-cue pedestrian detection and tracking from a moving vehicle. Int. Journal on Computer Vision **73**(1), 41–59 (2007)
115. Gavrila, D., Philomin, V.: Real-time object detection using distance transforms. In: IEEE Intelligent Vehicles Symposium. Stuttgart, Germany (1998)
116. Ge, J., Luo, Y., Tei, G.: Real-time pedestrian detection and tracking at nighttime for driver assistance systems. IEEE Trans. on Intelligent Transportation Systems **10**(2), 283–298 (2009)
117. Geiger, A., Lenz, P., Urtasun, R.: Are we ready for autonomous driving? In: IEEE Conf. on Computer Vision and Pattern Recognition. Providence, RI, USA (2012)
118. Geiger, A., Roser, M., Urtasun, R.: Efficient large-scale stereo matching. In: Asian Conf. on Computer Vision. Queenstown, New Zealand (2010)
119. Gerónimo, D.: A global approach to vision-based pedestrian detection for advanced driver assistance systems. PhD Thesis, Computer Vision Center, Universitat Autònoma de Barcelona (2010)
120. Gerónimo, D., López, A., Ponsa, D., Sappa, A.: Haar wavelets and edge orientation histograms for on board pedestrian detection. In: IAPR Iberian Conf. on Pattern Recognition and Image Analysis. Girona, Spain (2007)
121. Gerónimo, D., López, A., Sappa, A., Graf, T.: Survey of pedestrian detection for advanced driver assistance systems. IEEE Trans. on Pattern Analysis and, Machine Intelligence **32**(7), 1239–1258 (2010)
122. Gerónimo, D., Sappa, A., López, A., Ponsa, D.: Adaptive image sampling and windows classification for on-board pedestrian detection. In: Int. Conf. on Computer Vision Systems. Bielefeld, Germany (2007)
123. Gerónimo, D., Sappa, A., Ponsa, D., López, A.: 2D–3D-based on-board pedestrian detection system. Computer Vision and Image Understanding **114**(5), 583–595 (2010)
124. Giebel, J., Gavrila, D., Schnörr, C.: A Bayesian framework for multi-cue 3D object tracking. In: European Conf. on Computer Vision. Prague, Czech Republic (2004)
125. Girshick, R., Felzenszwalb, P., McAllester, D.: Object detection with grammar models. In: Advances in Neural Information Processing Systems. Granada, Spain (2011)
126. Goto, K., Kidono, K., Kimura, Y., Naito, T.: Pedestrian detection and direction estimation by cascade detector with multi-classifiers utilizing feature interaction descriptor. In: IEEE Intelligent Vehicles Symposium. Baden-Baden, Germany (2011)
127. Gould, S., Fulton, R., Koller, D.: Decomposing a scene into geometric and semantically consistent regions. In: Int. Conf. on Computer Vision. Kyoto, Japan (2009)
128. Graf, T., Seifert, K., Meinecke, M., Schmidt, R.: Human factors in designing advanced night vision systems. In: Congress and Exhibition on Intelligent Transport Systems and Services. Hannover, Germany (2005)
129. Gressmann, M., Palm, G., Löhlein, O.: Progressive pedestrian localization using neural networks. In: Intelligent Systems: Models and Applications, pp. 319–339. Springer (2013)
130. Grubb, G., Zelinsky, A., Nilsson, L., Rilbe, M.: 3D vision sensing for improved pedestrian safety. In: IEEE Intelligent Vehicles Symposium. Parma, Italy (2004)
131. Gualdi, G., Prati, A., Cucchiara, R.: A multi-stage pedestrian detection using monolithic classifiers. In: Advanced Video and Signal-Based Surveillance. Klagenfurt, Austria (2011)
132. Guo, L., Ge, P., Zhang, M., Li, L., Zhao, Y.: Pedestrian detection for intelligent transportation systems combining AdaBoost algorithm and support vector machine. Expert systems with applications **39**(4), 4274–4286 (2012)
133. Haar, A.: Zur theorie der orthogonalen funktionensysteme. Mathematische Annalen **69**(3), 331–371 (1910)

134. Halimeh, J., Roser, M.: Raindrop detection on car windshields using geometric–photometric environment construction and intensity-based correlation. In: IEEE Intelligent Vehicles Symposium. Xi'an, China (2009)

135. Haralick, R., Shanmugam, K., Dinstein, I.: Texture features for image classification. IEEE Trans. on Systems, Man, and, Cybernetics **3**(6), 610–621 (1973)

136. Rauch et al., The future of driving. Deliverable D32.1. Report on driver assessment methodology.

137. Heinrich, T.: Bewertung von technischen Maßnahmen zum Fußgängerschutz am Kraftfahrzeug. Tech. rep., Technische Universität Berlin, Studienarbeit, Berlin (2003)

138. Hinterstoisser, S., Lepetit, V., Ilic, S., Fua, P., Navab, N.: Dominant orientation templates for real-time detection of texture-less objects. In: IEEE Conf. on Computer Vision and Pattern Recognition. San Francisco, CA, USA (2010)

139. Hiromoto, M., Miyamoto, R.: Cascade classifier using divided cohog features for rapid pedestrian detection. In: Int. Conf. on Computer Vision Systems. Liege, Belgium (2009)

140. Hogg, D.: Model-based vision: a program to see a walking person. Image and Vision Computing **1**(1), 5–20 (1983)

141. Hoiem, D., Efros, A., Hebert, M.: Geometric context from a single image. In: Int. Conf. on Computer Vision. Beijing, China (2005)

142. Hoiem, D., Efros, A., Hebert, M.: Putting objects in perspective. In: IEEE Conf. on Computer Vision and Pattern Recognition. New York, NY, USA (2006)

143. Horbert, E., Rematas, K., Leibe, B.: Level-set person segmentation and tracking with multiregion appearance models and top-down shape information. In: Int. Conf. on Computer Vision. Barcelona, Spain (2011)

144. Hou, C., Ai, H., Lao, S.: Multiview pedestrian detection based on vector boosting. In: Asian Conf. on Computer Vision. Tokyo, Japan (2007)

145. Hu, Z., Uchimura, K.: U-V-disparity: an efficient algorithm for stereovision based scene analysis. In: IEEE Intelligent Vehicles Symposium. Las Vegas, NV, USA (2005)

146. Huang, C., Ai, H., Li, Y., Lao, S.: Vector boosting for rotation invariant multi-view face detection. In: Int. Conf. on Computer Vision. Beijing, China (2005)

147. Huang, C., Nevatia, R.: High performance object detection by collaborative learning of joint ranking of granules features. In: IEEE Conf. on Computer Vision and Pattern Recognition. San Francisco, CA, USA (2010)

148. Hubel, D., Wiesel, T.: Receptive fields, binocular interaction and functional architecture in cat's visual cortex. Journal of Physiology (London) **160**(1), 106–154 (1962)

149. Hubel, D., Wiesel, T.: Receptive fields and functional architecture in two nonstriate visual area (18 and 19) of the cat. Journal of Neurophysiology **28**(2), 229–289 (1965)

150. Hubel, D., Wiesel, T.: Receptive fields and functional architecture of monkey striate cortex. Journal of Physiology (London) **195**(1), 215–243 (1968)

151. Hubel, D., Wiesel, T.: Functional architecture of macaque monkey visual cortex. Proceedings of the Royal Society of London—Series B, Biological Sciences **198**(1130), 1–59 (1977)

152. Hussein, M., Porikli, F., Davis, L.: A comprehensive evaluation framework and a comparative study for human detectors. IEEE Trans. on Intelligent Transportation Systems **10**(3), 417–427 (2009)

153. Ioe, S., Forsyth, D.: Probabilistic methods for finding people. Int. Journal on Computer Vision **43**(1), 45–68 (2001)

154. Isard, M., Blake, A.: Contour tracking by stochastic propagation of conditional density. In: European Conf. on Computer Vision. Copenhagen, Denmark (1996)

155. Ito, S., Kubota, S.: Object classification using heterogeneous co-occurrence features. In: European Conf. on Computer Vision. Crete, Greece (2010)

156. Itti, L., Koch, C., Niebur, E.: A model of saliency-based visual attention for rapid scene analysis. IEEE Trans. on Pattern Analysis and, Machine Intelligence **20**(11), 1254–1259 (1998)

157. IVsource: Finally! adaptive cruise control arrives in the USA (2000)

158. IVsource: Iteris' lane departure warning system now available on mercedes trucks in Europe (2000)
159. Jain, A., Thormählen, T., Seidel, H., Theobalt, C.: MovieReshape: tracking and reshaping of humans in videos. In: ACM SIGGRAPHASIA Conf. and Exhib. on Computer Graphics and Interactive Techniques in Asia. Seoul, South Korea (2010)
160. Jones, M., Snow, D.: Pedestrian detection using boosted features over many frames. In: IEEE Conf. on Computer Vision and Pattern Recognition. Anchorage, AK, USA (2008)
161. Källhammer, J., Smith, K.: Assessing contextual factors that influence acceptance of pedestrian alerts by a night vision system. Human Factors: The Journal of the Human Factors and Ergonomics Society **54**(4), 654–662 (2012)
162. Kavukcuoglu, K., Sermanet, P., Boureau, Y., Gregor, K., Mathieu, M., LeCun, Y.: Learning convolutional feature hierachies for visual recognition. In: Advances in Neural Information Processing Systems. Vancouver, BC, Canada (2010)
163. Ke, Y., Sukthankar, R., Hebert, M.: Efficient visual event detection using volumetric features. In: Int. Conf. on Computer Vision. Beijing, China (2005)
164. Keller, C., Dang, T., Fritz, H., Joos, A., Rabe, C., Gavrila, D.: Active pedestrian safety by automatic braking and evasive steering. IEEE Trans. on Intelligent Transportation Systems **12**(4), 1292–1304 (2011)
165. Keller, C., Enzweiler, M., Gavrila, D.: A new benchmark for stereo-based pedestrian detection. In: IEEE Intelligent Vehicles Symposium. Baden-Baden, Germany (2011)
166. Keller, C., Enzweiler, M., Rohrbach, M., Fernández, D., Schnörr, C., Gavrila, D.: The benefits of dense stereo for pedestrian detection. IEEE Trans. on Intelligent Transportation Systems **12**(4), 1096–1106 (2011)
167. Keller, C., Hermes, C., Gavrila, D.: Will the pedestrian cross?: probabilistic path prediction based on learned motion features. In: DAGM Symposium. Frankfurt, Germany (2011)
168. Kittler, J., Hatef, M., Duin, R., Matas, J.: On combining classifiers. IEEE Trans. on Pattern Analysis and, Machine Intelligence **20**(3), 226–239 (1998)
169. Knoll, P.: HDR vision for driver assistance. In: B. Hoefflinger (ed.) High-Dynamic-Range (HDR) Vision, vol. 26, pp. 123–136. Springer Berlin Heidelberg (2007)
170. Krotosky, S., Trivedi, M.: Multimodal stereo image registration for pedestrian detection. In: IEEE Int. Conf. on Intelligent Transportation Systems. Seattle, WA, USA (2007)
171. Krotosky, S., Trivedi, M.: On color-, infrared-, and multimodal-stereo approaches to pedestrian detection. IEEE Trans. on Intelligent Transportation Systems **8**(4), 619–629 (2007)
172. Kuncheva, L.: Combining Pattern Classifiers. Methods and Algorithms. Wiley-Interscience (2004)
173. Labayrade, R., Aubert, D., Tarel, J.: Real time obstacle detection in stereovision on non flat road geometry through "v-disparity" representation. In: IEEE Intelligent Vehicles Symposium. Versailles, France (2002)
174. Lampert, C., Blashko, M., Hofmann, T.: Beyond sliding windows: Object localization by efficient subwindow search. In: IEEE Conf. on Computer Vision and Pattern Recognition. Anchorage, AK, USA (2008)
175. Laptev, I.: Improving object detection with boosted histograms. Image and Vision Computing **27**(5), 535–544 (2009)
176. Lazebnik, S., Schmidt, C., Ponce, J.: Beyond bags of features: Spatial pyramid matching for recognizing natural scene categories. In: IEEE Conf. on Computer Vision and Pattern Recognition. New York, NY, USA (2006)
177. LeCun, Y., Bengio, Y.: Convolutional networks for images, speech, and time-series. In: M. Arbib (ed.) The Handbook of Brain Theory and Neural Networks, pp. 255–258. MIT Press Cambridge (1995)
178. LeCun, Y., Boser, B., Denker, J., Henderson, D., Howard, R., Hubbard, W., Jackel, L.: Handwritten digit recognition with a back-propagation network. In: Advances in Neural Information Processing Systems. Denver, Colorado, USA (1989)
179. LeCun, Y., Bottou, L., Bengio, Y., Haffner, P.: Gradient-based learning applied to document recognition. Proceedings of the IEEE **86**(11), 2278–2324 (1998)

180. Leibe, B., Cornelis, N., Cornelis, K., Van Gool, L.: Dynamic 3D scene analysis from a moving vehicle. In: IEEE Conf. on Computer Vision and Pattern Recognition. Minneapolis, MN, USA (2007)
181. Leibe, B., Leonardis, A., Schiele, B.: Combined object categorization and segmentation with an implicit shape model. In: European Conf. on Computer Vision—Workshop on Statistical Learning in Computer Vision. Prague, Czech Republic (2004)
182. Leibe, B., Leonardis, A., Schiele, B.: Robust object detection with interleaved categorization and segmentation. Int. Journal on Computer Vision **77**(1–3), 259–289 (2008)
183. Leibe, B., Schiele, B.: Interleaved object categorization and segmentation. In: British Machine Vision Conference. Norwich, UK (2003)
184. Leibe, B., Schindler, K., Van Gool, L.: Coupled detection and trajectory estimation for multi-object tracking. In: Int. Conf. on Computer Vision. Rio de Janeiro, Brazil (2007)
185. Leibe, B., Seemann, E., Schiele, B.: Pedestrian detection in crowded scenes. In: IEEE Conf. on Computer Vision and Pattern Recognition. Washington, DC, USA (2005)
186. Levi, K., Weiss, Y.: Learning object detection from a small number of examples: the importance of good features. In: IEEE Conf. on Computer Vision and Pattern Recognition. Washington, DC, USA (2004)
187. Li, B., Yao, Q., Wang, K.: A review on vision-based pedestrian detection in intelligent transportation systems. In: IEEE Int. Conf. on Intelligent Transportation Systems. Anchorage, AK, USA (2012)
188. Lienhart, R., Maydt, J.: An extended set of Haar-like features for rapid object detection. In: IEEE Int. Conf. on Image Processing (2002)
189. Lin, Z., Davis, L.: A pose-invariant descriptor for human detection and segmentation. In: European Conf. on Computer Vision. Marseille, France (2008)
190. Lin, Z., Davis, L.: Shape-based human detection and segmentation via hierarchical part-template matching. IEEE Trans. on Pattern Analysis and, Machine Intelligence **32**(4), 604–618 (2010)
191. Lin, Z., Davis, L., Doermann, D., DeMenthon, D.: Hierarchical part-template matching for human detection and segmentation. In: Int. Conf. on Computer Vision. Rio de Janeiro, Brazil (2007)
192. Llorca, D., Sotelo, M., Hellín, A., Orellana, A., Gavilán, M., Daza, I., Llorente, A.: Stereo regions-of-interest selection for pedestrian protection: a survey. Transportation Research Part C: Emerging Technologies **25**, 226–237 (2012)
193. López, A., Hilgenstock, J., Busse, A., Baldrich, R., Lumbreras, F., Serrat, J.: Nighttime vehicle detection for intelligent headlight control. In: Advanced Concepts for Intelligent Vision Systems. Juan-les-Pins, France (2008)
194. Lowe, D.: Distinctive image features from scale-invariant keypoints. Int. Journal on Computer Vision **60**(2), 91–110 (2004)
195. The PASCAL Visual Object Classes Challenge 2012 (VOC2012) Results. http://www.pascal-network.org/challenges/VOC/voc2006/results.pdf
196. Mählisch, M., Oberländer, M., Löhlein, O., Gavrila, D., Ritter, W.: A multiple detector approach to low-resolution FIR pedestrian recognition. In: IEEE Intelligent Vehicles Symposium. Las Vegas, NV, USA (2005)
197. Maji, S., Berg, A., Malik, J.: Classification using intersection kernel support vector machines is efficient. In: IEEE Conf. on Computer Vision and Pattern Recognition. Anchorage, AK, USA (2008)
198. Marchal, P., Dehesa, M., Gavrila, D., Meinecke, M., Skellern, N., Viciguerra, R.: SAVE-U. final report. Tech. rep., Information Society Technology Programme of the EU (2005)
199. Marín, J., Gerónimo, D., Vázquez, D., López, A.: Pedestrian detection: Exploring virtual worlds. In: K. Hosny, J. Calleja (eds.) Handbook of Pattern Recognition: Methods and Applications. iConcept Press Ltd (2012)
200. Marín, J., Vázquez, D., Gerónimo, D., López, A.: Learning appearance in virtual scenarios for pedestrian detection. In: IEEE Conf. on Computer Vision and Pattern Recognition. San Francisco, CA, USA (2010)

201. Marr, D., Nishihara, H.: Representation and recognition of the spatial organization of three-dimensional shapes. Proceedings of the Royal Society of London—Series B, Biological Sciences **200**(1140), 269–294 (1978)
202. Marsi, S., Impoco, G., Ukovich, A., Carrato, S., Ramponi, G.: Video enhancement and dynamic range control of HDR sequences for automotive applications. EURASIP Journal on Advances in Signal Processing (2007)
203. Mateo, O., Otsuka, K.: Real-time visual tracker by stream processing. Journal of Signal Processing Systems **57**(2), 285–295 (2008)
204. Meis, U., Oberländer, M., Ritter, W.: Reinforcing the reliability of pedestrian detection in far-infrared sensing. In: IEEE Intelligent Vehicles Symposium, pp. 779–783. Parma, Italy (2004)
205. Messom, C., Barczak, A.: Stream processing for fast and efficient rotated haar-like features using rotated integral images. International Journal of Intelligent Systems Technologies and Applications **7**(1), 40–57 (2009)
206. Miau, F., Papageorgiou, C., Itti, L.: Neuromorphic algorithms for computer vision and attention. In: International Symposium on Optical Science and Technology, vol. 4479, pp. 12–23. Bellingham, WA, USA (2001)
207. Mikolajczyk, K., Schmid, C.: Indexing based on scale invariant interest points. In: Int. Conf. on Computer Vision. Vancouver, BC, Canada (2001)
208. Mikolajczyk, K., Schmid, C.: A performance evaluation of local descriptors. IEEE Trans. on Pattern Analysis and, Machine Intelligence **27**(10), 1615–1630 (2004)
209. Mikolajczyk, K., Schmid, C., Zisserman, A.: Human detection based on a probabilistic assembly of robust part detectors. In: European Conf. on Computer Vision. Prague, Czech Republic (2004)
210. Mita, T., Kaneko, T., Stenger, B., Hori, O.: Discriminative feature co-occurrence selection for object detection. IEEE Trans. on Pattern Analysis and, Machine Intelligence **30**(7) 1257–1269 (2008)
211. Mitsui, T., Fujiyoshi, H.: Object detection by joint features based on two-stage boosting. In: Int. Conf. on Computer Vision Worshop. Kyoto, Japan (2009)
212. Mitzel, D., Horbert, E., Ess, A., Leibe, B.: Multi-person tracking with sparse detection and continuous segmentation. In: European Conf. on Computer Vision. Crete, Greece (2010)
213. Mitzel, D., Leibe, B.: Taking mobile multi-object tracking to the next level: people, unknown objects, and carried items. In: European Conf. on Computer Vision. Firenze, Italy (2012)
214. Mitzel, D., Sudowe, P., Leibe, B.: Real-time multi-person tracking with time-constrained detection. In: British Machine Vision Conference. Dundee, UK (2011)
215. Moeslund, T., Hilton, A., Krüger, V.: A survey of advances in vision-based human motion capture and analysis. Computer Vision and Image Understanding **104**(2–3), 90–126 (2006)
216. Mohan, A., Papageorgiou, C., Poggio, T.: Example-based object detection in images by components. IEEE Trans. on Pattern Analysis and, Machine Intelligence **23**(4), 349–361 (2001)
217. Mohan, D.: Traffic safety and health in Indian cities. Journal of Transport Infrastructure **9**, 79–94 (2002)
218. Molineros, J., Cheng, S., Owechko, Y., Levi, D., Zhang, W.: Monocular rear-view obstacle detection using residual flow. In: European Conf. on Computer Vision—Workshop on Computer Vision in Vehicle Technology: From Earth to Mars at ECCV. Firenze, Italy (2012)
219. Mu, Y., Yan, S., Liu, Y., Huang, T., Zhou, B.: Discriminative local binary patterns for human detection in personal album. In: IEEE Conf. on Computer Vision and Pattern Recognition. Anchorage, AK, USA (2008)
220. Munder, S., Gavrila, D.: An experimental study on pedestrian classification. IEEE Trans. on Pattern Analysis and, Machine Intelligence **28**(11), 1863–1868 (2006)
221. Munder, S., Schnörr, C., Gavrila, D.: Pedestrian detection and tracking using mixture of view-based shape-texture models. IEEE Trans. on Intelligent Transportation Systems **9**(2), 333–343 (2007)
222. Nanda, H., Davis, L.: Probabilistic template based pedestrian detection in infrared videos. In: IEEE Intelligent Vehicles Symposium. Versailles, France (2002)

223. Nanni, L., Lumini, A.: Ensemble of multiple pedestrian representations. IEEE Trans. on Intelligent Transportation Systems 9(2), 365–369 (2008)
224. Nayar, S., Branzoi, V.: Adaptive dynamic range imaging: optical control of pixel exposures over space and time. In: Int. Conf. on Computer Vision. Nice, France (2003)
225. Nedevschi, S., Bota, S., Tomiuc, C.: Stereo-based pedestrian detection for collision-avoidance applications. IEEE Trans. on Intelligent Transportation Systems 20(3), 380–391 (2009)
226. Nedevschi, S., Danescu, R., Frentiu, D., Marita, T., Oniga, F., Pocol, C., Graf, T., Schmidt, R.: High accuracy stereovision approach for obstacle detection on non-planar roads. Proc. IEEE Intelligent Engineering Systems pp. 211–216 (2004)
227. Oberländer, M.: Hypermutation networks—a discrete approach to machine perception. In: Weightless Neural Networks Workshop. University of York, UK (2005)
228. Organisation internationale des constructeurs d'automobiles. http://www.oica.net
229. Ojala, T., Pietikäinen, M., Harwood, D.: A comparative study of texture measures with classification based on feature distributions. Pattern Recognition 29(1), 51–59 (1996)
230. Ojala, T., Pietikäinen, M., Mäenpää, T.: Multiresolution gray-scale and rotation invariant texture classification with local binary patterns. IEEE Trans. on Pattern Analysis and, Machine Intelligence 24(7), 971–987 (2002)
231. Oliveira, L., Nunes, U., Peixoto, P.: On exploration of classifier ensemble synergism in pedestrian detection. IEEE Trans. on Intelligent Transportation Systems 11(1), 16–27 (2010)
232. Oren, M., Papageorgiou, C., Sinha, P., Osuna, E., Poggio, T.: Pedestrian detection using wavelet templates. In: IEEE Conf. on Computer Vision and Pattern Recognition. San Juan, Puerto Rico (1997)
233. Organization, W.H.: World Health Statistics. World Health Organization, France (2008)
234. Ott, P., Everingham, M.: Implicit color segmentation features for pedestrian and object detection. In: Int. Conf. on Computer Vision. Kyoto, Japan (2009)
235. Ouyang, W., Wang, X.: A discriminatively deep model for pedestrian detection with occlusion handling. In: IEEE Conf. on Computer Vision and Pattern Recognition. Providence, RI, USA (2012)
236. Overett, G., Petersson, L., Brewer, N., Andersson, L., Pettersson, N.: A new pedestrian dataset for supervised learning. In: IEEE Intelligent Vehicles Symposium. Eindhoven, The Netherlands (2008)
237. Paisitkriangkrai, S., Shen, C., Ziang, J.: Fast pedestrian detection using a cascade of boosted covariance features. IEEE Trans. on Circuits and Systems for Video Technology 18(8), 1140–1150 (2008)
238. Pang, Y., Yan, H., Yuan, Y., Wang, K.: Robust CoHOG feature extraction in human-centered image/video management system. IEEE Trans. on Systems, Man, and Cybernetics—Part B 42(2), 458–468 (2012)
239. Pang, Y., Yuan, Y., Li, X., Pan, J.: Efficient HOG human detection. Signal Processing 91(4), 773–781 (2011)
240. Papageorgiou, C., Oren, M., Poggio, T.: A general framework for object detection. In: Int. Conf. on Computer Vision. Bombay, India (1998)
241. Papageorgiou, C., Poggio, T.: A trainable system for object detection. Int. Journal on Computer Vision 38(1), 15–33 (2000)
242. Parikh, D., Zitnick, C.: Finding the weakest link in person detectors. In: IEEE Conf. on Computer Vision and Pattern Recognition. Colorado Springs, CO, USA (2011)
243. Park, D., Ramanan, D., Fowlkes, C.: Multiresolution models for object detection. In: European Conf. on Computer Vision. Crete, Greece (2010)
244. Park, S., Trivedi, M.: Driver activity analysis for intelligent vehicles: Issues and development framework. In: IEEE Intelligent Vehicles Symposium. Las Vegas, NV, USA (2005)
245. Parra, I., Fernández, D., Sotelo, M., Bergasa, L., Revenga, P., Nuevo, J., Ocaña, M., García, M.: Combination of feature extraction method for SVM pedestrian detection. IEEE Trans. on Intelligent Transportation Systems 8(2), 292–307 (2007)
246. Pedersoli, M., Vedaldi, A., Gonzàlez, J.: A coarse-to-fine approach for fast deformable object detection. In: IEEE Conf. on Computer Vision and Pattern Recognition. Colorado Springs, CO, USA (2011)

247. Pfeiffer, D., Erbs, F., Franke, U.: Pixels, stixels, and objects. In: European Conf. on Computer Vision—Workshop on Computer Vision in Vehicle Technology: From Earth to Mars at ECCV. Firenze, Italy (2012)
248. Pfeiffer, D., Franke, U.: Efficient representation of traffic scenes by means of dynamic stixels. In: IEEE Intelligent Vehicles Symposium. San Diego, CA, USA (2010)
249. Pfeiffer, D., Franke, U.: Towards a global optimal multi-layer stixel representation of dense 3d data. In: British Machine Vision Conference. Dundee, UK (2011)
250. Philomin, V., Duraiswami, R., Davis, L.: Pedestrian tracking from a moving vehicle. In: IEEE Intelligent Vehicles Symposium. Dearborn, MI, USA (2000)
251. Pietikäinen, M., Hadid, A., Zhao, G., Ahonen, T.: Computer Vision Using Local Binary Patterns. Springer (2011)
252. Pishchulin, L., Jain, A., Andrilika, M., Thormählen, T., Schiele, B.: Articulated people detection and pose estimation: Reshaping the future. In: IEEE Conf. on Computer Vision and Pattern Recognition. Providence, RI, USA (2012)
253. Pishchulin, L., Jain, A., Wojek, C., Andrilika, M., Thormählen, T., Schiele, B.: Learning people detection models from few training samples. In: IEEE Conf. on Computer Vision and Pattern Recognition. Colorado Springs, CO, USA (2011)
254. Ponsa, D., López, A., Serrat, J., Lumbreras, F., Graf, T.: 3D vehicle sensor based on monocular vision. In: IEEE Int. Conf. on Intelligent Transportation Systems. Vienna, Austria (2005)
255. Pouli, T., Cunningham, D., Reinhard, E.: Image statistics and their applications in computer graphics. In: European Computer Graphics Conference and Exhibition. Norrköping, Sweden (2010)
256. PREVENT. http://www.prevent-ip.org
257. Prisacariu, V., Reid, I.: fasthog—a real-time gpu implementation of hog. Tech. Rep. 2310/09, Department of Engineering Science, Oxford University (2009)
258. Amazon Mechanical Turk. http://www.mturk.com
259. Dariu M. Gavrila: Smart Cars for Safe Pedestrians (IV'12). http://www.youtube.com/watch?v=esIwzZN1skE
260. Quiñonero-Candela, J., Sugiyama, M., Schwaighofer, A., Lawrence, N. (eds.): Dataset shift in machine learning. Neural Information Processing. The MIT Press (2008)
261. Rabe, C., Franke, U., Gehrig, S.: Fast detection of moving objects in complex scenarios. In: IEEE Intelligent Vehicles Symposium. Istanbul, Turkey (2007)
262. Ramanan, D.: Using segmentation to verify object hypotheses. In: IEEE Conf. on Computer Vision and Pattern Recognition. Minneapolis, MI, USA (2007)
263. Rohrbach, M., Enzweiler, M., Gavrila, D.: High-level fusion of depth and intensity for pedestrian classification. In: DAGM Symposium. Jena, Germany (2009)
264. Ross, J., Irani, L., Silberman, M., Zaldivar, A., Tomlinson, B.: Who are the crowdworkers? shifting demographics in Mechanical Turk. In: ACM Int. Conf. on Human Factors in Computing Systems. Atlanta, GA, USA (2010)
265. Russell, B., Torralba, A., Murphy, K., Freeman, W.: LabelMe: a database and web-based tootl for image annotation. Int. Journal on Computer Vision 77(1–3), 157–173 (2008)
266. Sabzmeydani, P., Mori, G.: Detecting pedestrians by learning shapelet features. In: IEEE Conf. on Computer Vision and Pattern Recognition. Minneapolis, MN, USA (2007)
267. Sappa, A., Dornaika, F., Ponsa, D., Gerónimo, D., López, A.: An efficient approach to onboard stereo vision system pose estimation. IEEE Trans. on Intelligent Transportation Systems 9(3), 476–490 (2008)
268. http://www.save-u.org/
269. Scharstein, D., Szeliski, R.: A taxonomy and evaluation of dense two-frame stereo correspondence algorithms. Int. Journal on Computer Vision 47(1–3), 7–42 (2002)
270. Schwartz, W., Kembhavi, A., Harwood, D., Davis, L.: Human detection using partial least squares analysis. In: Int. Conf. on Computer Vision. Kyoto, Japan (2009)
271. Seemann, E., Leibe, B., Mikolajczyk, K., Schiele, B.: An evaluation of local shape-based features for pedestrian detection. In: British Machine Vision Conference. Oxford, UK (2005)

272. Seemann, E., Leibe, B., Schiele, B.: Multi-aspect detection of articulated objects. In: IEEE Conf. on Computer Vision and Pattern Recognition. New York, NY, USA (2006)
273. Seemann, E., Schiele, B.: Cross-articulation learning of robust detection of pedestrians. In: DAGM Symposium. Berlin, Germany (2006)
274. Sermanet, P., Kavukcuoglu, K., LeCun, Y.: Traffic sign recognition with multi-scale convolutional networks. In: IEEE Int. Joint Conf. on Neural Networks. San Jose, CA, USA (2011)
275. Shao, W.: Animating autonomous pedestrians. PhD Thesis, Dept. C. S., Courant Inst. of Mathematical Sciences, New York Univesity (2006)
276. Shashua, A., Gdalyahu, Y., Hayun, G.: Pedestrian detection for driving assistance systems: Single-frame classification and system level performance. In: IEEE Intelligent Vehicles Symposium. Parma, Italy (2004)
277. Shimizu, H., Poggio, T.: Direction estimation of pedestrian from images. Tech. rep., Artificial Intelligence Group, Massachusetts Institute of Technology (2003)
278. Singh, V., Wu, B., Nevatia, R.: Pedestrian tracking by associating tracklets using detection residuals. In: Workshop on Motion and Video Computing. Copper Mountain, CO, USA (2008)
279. Smith, S., Brady, J.: Susan—a new approach to low level image processing. Int. Journal on Computer Vision 23(1), 45–78 (1997)
280. Government of India—Department of road transport and highways. Total number of motor vehicles in India 1951–2004 (2004)
281. National Highway Traffic Safety Administration—National Center for Statistics and Analysis, Traffic Safety Facts (2007)
282. Sturgess, P., Alahari, K., Ladicky, L., Torr, P.: Combining appearance and structure from motion features for road scene understanding. In: British Machine Vision Conference. London, UK (2009)
283. Suard, F., Rakotomamonjy, A., Bensrhair, A., Broggi, A.: Pedestrian detection using infrared images and histograms of oriented gradients. In: IEEE Intelligent Vehicles Symposium. Tokyo, Japan (2006)
284. Sudowe, P., Leibe, B.: Efficient use of geometric constraints for sliding-window object detection in video. In: Int. Conf. on Computer Vision Systems. Sophia Antipolis, France (2011)
285. Sun, H., Wang, C., Wang, B., El-Sheimy, N.: Pyramid binary pattern features for real-time pedestrian detection from infrared videos. Neurocomputing 74(5), 797–804 (2011)
286. Sun, Z., Bebis, G., Miller, R.: On-road vehicle detection: a review. IEEE Trans. on Pattern Analysis and, Machine Intelligence 28(5), 694–711 (2006)
287. Szarvas, M., Yoshizawa, A., Yamamoto, M., Ogata, J.: Pedestrian detection with convolutional neural networks. In: IEEE Intelligent Vehicles Symposium. Las Vegas, NV, USA (2005)
288. Tang, D., Liu, Y., Kim, T.: Fast pedestrian detection by cascade random forest with dominant orientation. In: British Machine Vision Conference. Surrey, UK (2012)
289. Tang, S., Andriluka, M., Schiele, B.: Detection and tracking of occluded people. In: British Machine Vision Conference. Surrey, UK (2012)
290. http://www.tass-safe.com
291. Tian, Q., Sun, H., Luo, Y., Hu, D.: Nighttime pedestrian detection with a normal camera using SVM classifier. In: Int. symp. on neural networks (2005)
292. Torralba, A., Efros, A.: Unbiased look at dataset bias. In: IEEE Conf. on Computer Vision and Pattern Recognition. Colorado Springs, CO, USA (2011)
293. Tosato, D., Farenzena, M., Cristani, M., Murino, V.: Part-based human detection on Riemannian manifolds. In: Int. Conf. in Pattern Recognition. Istanbul, Turkey (2010)
294. Tosato, D., Farenzena, M., Cristani, M., Murino, V.: A re-evaluation of pedestrian detection on Riemannian manifolds. In: Int. Conf. in Pattern Recognition. Istanbul, Turkey (2010)
295. Tran, D., Forsyth, D.: Configuration estimates improve pedestrian finding. In: Advances in Neural Information Processing Systems. Vancouver, Canada (2007)
296. Trivedi, M., Gandhi, T., McCall, J.: Looking-in and looking-out of a vehicle: computer-vision-based enhanced vehicle safety. IEEE Trans. on Intelligent Transportation Systems 8(1), 108–120 (2007)

297. Tsuji, T., Hattori, H., Watanabe, M., Nagaoka, N.: Development of night-vision system. IEEE Trans. on Intelligent Transportation Systems **3**(3), 203–209 (2002)
298. Tuzel, O., Porikli, F., Meer, P.: Region covariance: a fast descriptor for detection and classification. In: European Conf. on Computer Vision. Graz, Austria (2006)
299. Tuzel, O., Porikli, F., Meer, P.: Human detection via classification on Riemannian manifolds. In: IEEE Conf. on Computer Vision and Pattern Recognition. Minneapolis, MN, USA (2007)
300. Tuzel, O., Porikli, F., Meer, P.: Pedestrian detection via classification on Riemannian manifolds. IEEE Trans. on Pattern Analysis and, Machine Intelligence **30**(10), 1–15 (2008)
301. Unger, C., Wahl, E., Ilic, S.: Efficient stereo matching for moving cameras and decalibrated rigs. In: IEEE Intelligent Vehicles Symposium. Baden-Baden, Germany (2011)
302. van der Mark, W., Gavrila, D.: Real-time dense stereo for intelligent vehicles. IEEE Trans. on Intelligent Transportation Systems **7**(1), 38–50 (2006)
303. Vapnik, V.: The Nature of Statistical Learning Theory. Springer (1995)
304. Vavilin, A., Deb, K., Kim, T., Jo, K.: Road sign detection method based on fast hdr image generation technique. In: IEEE Int. Conf. on Image and Vision Computing New Zealand. Queenstown, New Zealand (2010)
305. Vázquez, D., López, A., Ponsa, D.: Unsupervised domain adaptation of virtual and real worlds for pedestrian detection. In: Int. Conf. in Pattern Recognition. Tsukuba, Japan (2012)
306. Vázquez, D., López, A., Ponsa, D., Gerónimo, D.: Interactive training of human detectors. In: A. Sappa, J. Vitrià (eds.) Multimodal Interaction in Pattern Recognition and Computer Vision, pp. 169–184. Springer (2013)
307. Vázquez, D., López, A., Ponsa, D., Marín, J.: Cool world: domain adaptation of virtual and real worlds for human detection using active learning. In: Advances in Neural Information Processing Systems—Workshop on Domain Adaptation: Theory and Applications. Granada, Spain (2011)
308. Vázquez, D., López, A., Ponsa, D., Marín, J.: Virtual worlds and active learning for human detection. In: ACM Int. Conf. on Multimodal Interaction. Alicante, Spain (2011)
309. Viola, P., Jones, M.: Rapid object detection using a boosted cascade of simple features. In: IEEE Conf. on Computer Vision and Pattern Recognition. Kauai, HI, USA (2001)
310. Viola, P., Jones, M., Snow, D.: Detecting pedestrians using patterns of motion and appearance. In: Int. Conf. on Computer Vision. Nice, France (2003)
311. Vlacic, L., Parent, M., Harashima, F.: Intelligent Vehicle Technologies. Butterworth-Heinemann (2001)
312. Walk, S., Majer, N., Schindler, K., Schiele, B.: New features and insights for pedestrian detection. In: IEEE Conf. on Computer Vision and Pattern Recognition. San Francisco, CA, USA (2010)
313. Wang, L., Shi, J., Song, G., Shen, I.: Object detection combining recognition and segmentation. In: Asian Conf. on Computer Vision. Osaka, Japan (2007)
314. Wang, Q., Pang, J., Qin, L., Jiang, S., Huang, Q.: Justifying the importance of color cues in object detection: a case study on pedestrian. In: J. Jin, C. Xu, M. Xu (eds.) The Era of Interactive Media, pp. 387–397. Springer (2013)
315. Wang, X., Han, T., Yan, S.: An HOG-LBP human detector with partial occlusion handling. In: Int. Conf. on Computer Vision. Kyoto, Japan (2009)
316. Watanabe, T., Ito, S., Yoki, K.: Co-occurrence histograms of oriented gradients for pedestrian detection. In: Pacific-Rim Symposium on Image and Video Technology. Tokyo, Japan (2009)
317. Wei, X., Phung, S., Bouzerdoum, A.: Pedestrian sensing using time-of-flight range camera. In: IEEE Conf. on Computer Vision and Pattern Recognition. Colorado Springs, CO, USA (2011)
318. Welch, G., Bishop, G.: An introduction to the Kalman filter. Tech. rep., University of North Carolina at Chapel Hill, Department of Computer Science (2002)
319. Wöhler, C., Anlauf, J.: An adaptable time-delay neural-network algorithm for image sequence analysis. IEEE Trans. on Neural Networks **10**(6), 1531–1536 (1999)
320. Wöhler, C., Anlauf, J., Pörtner, T., Franke, U.: A time-delay neural-network algorithm for real-time pedestrian recognition. In: IEEE Int. Conf. on Intelligent Vehicles. Stuttgart, Germany (1998)

321. Wojek, C., Dorkó, G., Schulz, A., Schiele, B.: Sliding-windows for rapid object class localization: a parallel technique. In: Symposium of the German Association for Pattern Recognition. Munich, Germany (2008)
322. Wojek, C., Schiele, B.: A performance evaluation of single and multi-feature people detection. In: DAGM Symposium. Munich, Germany (2008)
323. Wojek, C., Walk, S., Roth, S., Schiele, B.: Monoclar 3D scene understanding with explicit occlusion reasoning. In: IEEE Conf. on Computer Vision and Pattern Recognition. Colorado Springs, CL, USA (2011)
324. Wojek, C., Walk, S., Roth, S., Schindler, K., Schiele, B.: Monocular visual scene understanding: understanding multi-object traffic scenes. IEEE Trans. on Pattern Analysis and Machine Intelligence (in press) (2013)
325. Wojek, C., Walk, S., Schiele, B.: Multi-cue onboard pedestrian detection. In: IEEE Conf. on Computer Vision and Pattern Recognition. Miami, FL, USA (2009)
326. Wu, B., Nevatia, R.: Detection of multiple, partially occluded humans in a single image by Bayesian combination of edgelet part detection. In: Int. Conf. on Computer Vision. Beijing, China (2005)
327. Wu, B., Nevatia, R.: Cluster boosted tree classifier for multi-view, multi-pose object detection. In: Int. Conf. on Computer Vision. Rio de Janeiro, Brazil (2007)
328. Wu, B., Nevatia, R.: Detection and tracking of multiple, partially occluded humans by Bayesian combination of edgelet part detectors. Int. Journal on Computer Vision $75(2)$, 247–266 (2007)
329. Wu, B., Nevatia, R.: Simultaneous object detection and segmentation by boosting local shape feature based classifier. In: IEEE Conf. on Computer Vision and Pattern Recognition. Minneapolis, MN, USA (2007)
330. Xu, R., Jiao, J., Zhang, B., Ye, Q.: Pedestrian detection in images via cascaded l1-norm minimization learning method. Pattern Recognition $45(7)$, 2573–2583 (2012)
331. Xu, Y., Cao, X., Qiao, H.: An efficient tree classifier ensemble approach for pedestria detection. IEEE Trans. on Systems, Man, and Cybernetics—Part B $41(1)$, 107–117 (2011)
332. Xu, Y., Xu, D., Lin, S., Han, T., Cao, X., Li, X.: Detection of sudden pedestrian crossings for driving assistance systems. IEEE Trans. on Systems, Man, and Cybernetics—Part B $42(3)$, 729–739 (2012)
333. Yamauchi, Y., Takaki, M., Yamashita, T., Fujiyoshi, H.: Feature co-occurrence representation based on boosting for object detection. In: IEEE Conf. on Computer Vision and Pattern Recognition Worshop. San Francisco, CA, USA (2010)
334. Ye, Q., Liang, J., Jiao, J.: Pedestrian detection in video images via error correcting output code classification of manifold subclasses. IEEE Trans. on Intelligent Transportation Systems $13(1)$, 193–202 (2012)
335. Yu, L., Yao, W., Liu, H., Liu, F.: A monocular vision based pedestrian detection system for intelligent vehicles. In: IEEE Intelligent Vehicles Symposium. Eindhoven, The Netherlands (2008)
336. Zhang, C., Viola, P.: Multiple-instance pruning for learning efficient cascade detectors. In: Advances in Neural Information Processing Systems. Vancouver, BC, Canada (2007)
337. Zhang, J., Huang, K., Yu, Y., Tan, T.: Boosted local structured HOG-LBP for object localization. In: IEEE Conf. on Computer Vision and Pattern Recognition. San Francisco, CA, USA (2010)
338. Zhang, L., Li, Y., Nevatia, R.: Global data association for multi-object tracking using network flows. In: IEEE Conf. on Computer Vision and Pattern Recognition. Anchorage, AK, USA (2008)
339. Zhang, L., Nevatia, R.: Efficient scan-window based object detection using GPGPU. In: IEEE Conf. on Computer Vision and Pattern Recognition. Anchorage, AK, USA (2008)
340. Zhang, W., Zelinsky, G., Samaras, D.: Real-time accurate object detection using multiple resolutions. In: Int. Conf. on Computer Vision. Rio de Janeiro, Brazil (2007)
341. Zhao, L., Thorpe, C.: Stereo and neural network-based pedestrian detection. IEEE Trans. on Intelligent Transportation Systems $1(3)$, 148–154 (2000)

342. Zhu, Q., Avidan, S., Yeh, M., Cheng, K.: Fast human detection using a cascade of histrograms of oriented gradients. In: IEEE Conf. on Computer Vision and Pattern Recognition. New York City, NY, USA (2006)
343. Zhu, X., Vondrick, C., Ramanan, D., Fowlkes, C.: Do we need more training data or better models for object detection? In: British Machine Vision Conference. Surrey, UK (2012)
344. Zucker, S., Terzopoulos, D.: Finding structure in co-occurrence matrices for texture analysis. Computer Graphics and Image Processing **12**, 286–308 (1980)